Mohamed Osman Akasha

Darfur

A tragedy of climate change

Anchor Academic
Publishing

Akasha, Mohamed Osman: Darfur: A tragedy of climate change.
Hamburg, Anchor Academic Publishing 2013

Buch-ISBN: 978-3-95489-095-8
PDF-eBook-ISBN: 978-3-95489-595-3
Druck/Herstellung: Anchor Academic Publishing, Hamburg, 2013

Bibliografische Information der Deutschen Nationalbibliothek:
Die Deutsche Nationalbibliothek verzeichnet diese Publikation in der Deutschen Nationalbibliografie; detaillierte bibliografische Daten sind im Internet über http://dnb.d-nb.de abrufbar.

Bibliographical Information of the German National Library:
The German National Library lists this publication in the German National Bibliography. Detailed bibliographic data can be found at: http://dnb.d-nb.de

Cover picture: United Nations Environmental Programme (UNEP)
http://www.opensocietyfoundations.org/sites/default/files/styles/mw_collection_910/public/photos/20051201-addario-mw11-001-910.jpg?itok=62XoOD3n

Hermannstal 119k, 22119 Hamburg
http://www.diplomica-verlag.de, Hamburg 2013
Printed in Germany

Table of contents

Acknowledgment

This book was possible because of the extensive support of my beloved wife who devoted herself for our family. I would also like to thank my lovely daughters for tolerating my absence patiently while I am writing this book. My sincere thanks go to my Parents. I do extend special gratitude to many individuals in all parts of Sudan and United Kingdom who helped me writing this book.

Summary

Climate change presents a serious threat to the security and prosperity of all countries. The effects of climate change and its security implications have now been at the forefront of international attention. Academic researches on environmental change and security gained popularity in political science and security studies in the 1990s. With Cold War-related security issues on the decline, policymakers began looking more closely at non-traditional security concerns such as environmental change, poverty and diseases. In so doing, the idea of what constitutes state security expanded beyond the risk of direct military aggression from hostile states to concerns about the regional instability that could affect economic security and draw governments into regional conflicts.

Few quantitative and qualitative studies conclude that, climate change in itself is unlikely to produce violent conflict, but rather, it could serve as a "threat multiplier" whereby environmental degradation caused by climate change may exacerbate many of the underlining causes linked to violent conflict. However this conclusion is widely generalized, since it has put all countries in one basket despite the numerous differences and variation between countries as far as adaptive mechanism and mitigation technologies are concerned. Using Darfur as a case study, this book examines the effects of climate change in poor or less developed countries, and critically analyzing the concept of climate change as a security threat that has ignited the conflict in Darfur, critically analyzing the role of drought, desertification, decreased rainfall, land degradation and migration in the conflict in Darfur. While acknowledging the prominent role of the International community in solving this crisis, this work intends to illustrate that, the absence of a common view on the nature and root causes of the conflict has hampered international convergence about how to act on Darfur crisis. This has delayed a coherent response and has contributed to the escalation of the conflict. Also the mischaracterization of the causes and nature of the conflict in Darfur has contributed to oversimplified views, which allowed the conflict to be politicized in a way that has complicated the search for solutions. This has hampered progress toward defining a political settlement, and toward finding a formula for allowing the various population groups to pursue compatible and sustainable livelihoods and as a result of the wrong diagnosis of the problem people of Darfur are still suffering. This books aims to draw the attention of international community, non-governmental organization and public to the reality of the root causes of the crisis in Darfur

Introduction

This book seeks to critically analyze the role of climate change in intrastate conflicts in less developed countries, and links between climate change and the untraditional concept of security threats. The book will also narrow in to analysis and takes the crisis in Darfur as a case study. It will cover the period when the climate change started to hit Darfur in 1970s passing through the period of 1980s when Darfur was devastated by the major famine. Then it will focus on the era when the conflict escalated in 2003 and the role of the international community till 2008.

Darfur started to attract the attention of the international community following the outbreak of the conflict in 2003 owing to the escalation of the violent conflict that led to massive displacements, killing and deaths.

Since 2003, much is being written on what is happening on the ground, much less about the root causes of the conflict and that is because it has been looked at from a political perspective rather than scientific one, and therefore it has been described by many activist groups, countries and humanitarian organization as a genocide resembling the tragedy of the 21st century. However, the root causes of the conflict have not been addressed properly. Few scholars and scientists shed lights on these root causes, but the important stakeholders ignored the fact that climate change with increasing desertification and decreased rainfall is a major factor behind the crises which had started since 1970s. The symptoms of climate change in Darfur intensified in 1980s when the region witnessed a severe drought and famine, as a result, more people are competing for access to land, water, and other natural resources in Darfur. In 2003 the conflict turned to be an armed conflict between the various tribes over natural resources, especially water and land. These impacts include expanding desertification, decreased rainfall and land degradation left dire consequences, as pastoralists (Abbala) have migrated south for improved grazing for their herds, yet farmers have denied them access due to their marginal lands. As a result, more Darfurians are competing for access to land, water, and other natural resources than at any other time. The increased competition only further aggravates the already uneasy political, social, and ethnic relationships in the Darfur region.

Map No. 3707 Rev. 7 UNITED NATIONS
May 2004

Department of Peacekeeping Operations
Cartographic Section

Map No. 4458 Rev.2 UNITED NATIONS
March 2012

Department of Field Support
Cartographic Section

The origins of the conflict in Darfur

People of Darfur have been devastated by war, and its aftermath has been sorrowful story of suffering, displacement and death. The war has become one of the most misrepresented and misunderstood conflicts in recent history. Analysts and activists have oversimplified the causes of the war, slighting its historical and systemic causes. For years, public commentators ignored the nature and the real cause of the violence in Darfur, causing important misperceptions among the public and in the policy community[1].

The region of Darfur had witnessed a severe drought since 1970s caused by decreased rainfall which led to land degradation; in addition to that, the population has significantly increased. The inhabitants of Darfur depend heavily on the natural resource base for their socio-economic activities. Land and water access are crucial for sustainable livelihoods. The majority of people earn their livelihoods through subsistence agriculture, either farming or pastoralist[2]. With the increasing competition for land and water resources and the lack of local conflict resolution mechanisms, the relations between the competing tribes have intensified and developed to an armed conflict with traditional weapons.

The conflict's origin goes back to land disputes between semi-nomadic livestock herders and those who practice sedentary agriculture[3], as nomadic tribes facing drought are going after the territory of sedentary farmers[4].

In 1980s Darfur was badly hit by unprecedented drought that caused the well-known famine in Darfur and the conflict between tribes continued to evolve as pastoralists have migrated south for improved grazing for their herds, yet farmers have denied them access

1. Foreign policy briefing, volume No, 89 June 1, 2010.
2. Fadul, A. A. *Natural Resources Management for Sustainable Peace in Darfur. Sudan, University for Peace, (2004)*, p. 34.
3. Scott Straus, *Darfur and the Genocide Debate*, Foreign Affairs magazine, Vol. 84, No. 1 (Jan. – Feb 2005), pp. 123/133.
4. Bechtold, P. K. *A History of Modern Sudan.* Middle East Journal 2009, 63(1), 149 – 150.

due to their marginal lands. The year 2003 witnessed the escalation of the current crises but this time light mechanical weapons were used and insurgency against the government was announced by some tribal groups alleging that the central government is intentionally ignoring the development of Darfur. Although the conflict in Darfur in its holistic is approaching its fourth decade, yet no lasting solution was found. Now it has been nine years since the armed conflict in Darfur attract the attention of the international community following the outbreak of the conflict in 2003, during this period many Security Council resolutions have been issued against Sudan. Many unilateral sanctions were imposed on Sudan, an economical sanction has been applied to Sudan by EU, and moreover, troops from UN have been sent to Darfur. Recently, the case has been transferred from the Security Council to the International Criminal Court[5], accordingly, warrant of arrest was issued against the Sudanese President and other officials, yet the crisis was not solved. If all these actions are unable to find the cure, then there is something wrong with the diagnosis of the problem.

The conflict is being presented in the media and by some stakeholders as a war between Arabs and Africans, with Arab militias, called *janjawid*, carrying out massacres, rape and pillage with the support of the Khartoum Government, and sometimes presented as a war between Muslim and non-Muslim, while this is far away from the fact, Darfur region is entirely inhabited by Muslim tribes and the usage of the terms "Arab" and "Black" has been opposed, because all parties involved in the Darfur conflict whether they are referred to as 'Arab' or as 'African,' are equally indigenous Muslim black African[6].

Most of the policy makers, activists and researchers ignored or even denied the climate change factor as the major cause of the conflict. Instead they started pushing the International community to take coercive actions against the government in Sudan, and that was an important factor that deprived the International community from looking into the root causes of the conflict.

5. Security Council Resolution 1593 (2005).
6. Alex de Waal & A. H. Abdel Salam eds, *The Phoenix State: Civil Society and the Future of Sudan*, (Red Sea Press 200)1, p.45
7. Failed states ranking. Available at > www.fundforpeace.org<, [accessed on 9 June /2012].

The scientific debate on climate change as a cause of conflicts

The idea that climate change and resource scarcity can lead to conflict has become a wide-spread assumption. Yet, a consensus has not been reached. Academic writings and scientific works on the issue differ in the final judgments. While there are strong theoretic arguments and qualitative case studies supporting the significant link between resource scarcity and armed conflict, some quantitative studies contradict these findings by believing in weak empirical evidence for such claims. This book would add a new dimension for this subject as it argues that the impacts of climate change cannot be the same in every society, the severity of these impacts depend on the situation within the state itself. For instance if a developed and less developed states have been hit by the same drought the poorest will be worst affected, While the developed state can absorb the consequences and adapt itself by using mitigation technology and the already existing adaptive capacities. In the case of Darfur, since 1990 Sudan has been ranked the third failed state in the world by many specialized organizations like fund for peace Institute[7], a country ranking based on a set of indicators assessing stability and vulnerability and that includes social, economic, and political indicators such as demographic pressures, refugee flows, uneven economic development or severe economic decline. The vulnerability of people in Darfur is shaped not only by the persistence of poverty, the lack of good infrastructure, the difficulty of getting a foothold in the world market, and thus the intractability of underdevelopment, but also by the effects of armed conflict which has started since August 1955 and continued under the ruling of the successive governments took power in Sudan.

The literature written on resource scarcity has focused around environmental degradation as a cause of conflict also known as the neo-Malthusian paradigm[8] or the Toronto group. Leading scholars in this field include Homer-Dixon, Baechler (1999), and Kahl (2006). As the founder of resource scarcity theory, Homer-Dixon identifies three forms of resource scarcity: "demand-induced scarcity", which results from population growth; "supply-induced scarcity", which results from the depletion or degradation of a resource; and "structural scarcity", which refers to the "distribution of the resource."[9] The interaction of these three factors is more likely to produce intrastate conflicts versus interstate wars. Furthermore, in the journal of International Security, Homer-Dixon introduces his model for identifying environment-conflict linkages in which he stated that, environmental change, population growth and unequal resource distribution would lead to internal conflicts .[10] The degradation and depletion of environmental resources is only one source of environmental scarcity, two other important sources are population growth and unequal resource distribution.[11] These main resources water, land, fisheries, and forests (renewable resources): are essential for livelihoods of millions of people, especially for subsistence users. With renewable resources, Homer-Dixon identifies two main patterns of interactions for environmental scarcity: resource capture and ecological marginalization.[12]

---

8. Homer-Dixon, T. F., 1994, 'Environmental Scarcities and Violent Conflict: Evidence from Cases', International Security, vol. 19, no. 1, pp. 5-40.
9. Gleditch, Nils Petter , Environmental Change, Security and Conflict. United States Institute of Peace Press 2007, pp.177-196; p. 179.
10. Homer-Dixon, Thomas. "Environmental Scarcities and Violent Conflict: Evidence from Cases," *International Security*, Vol. 19, No. 1 (summer, 1994), pp. 5-40; p. 10.
11. Ibid, p. 40.
12. Ibid, p. 10.

Resource capture occurs with population growth and natural resource degradation. Ecological marginalization occurs when environmental changes from degradation lead to significant social effects, such as human migration into more ecologically sensitive areas. Additionally, environmental degradation of natural resources has impacts across horizontal and vertical levels. Environmental degradation may cause countless often subtle changes in developing societies. If the state cannot respond to the environmental degradation, either through markets, adaptation, or mitigation policies, then any existing social/economic/political political cleavages can erupt into conflict. According to Kahl (2006), state failure and the state exploitation can cause violent conflict over scarce resources[13.] Agreeing with Homer Dixon; Kahl argues that resource scarcity, stemming from demographic and environmental stresses (DES), can place significant pressure on state institutions that may lack the capacity to manage the contested resource. At the field level, societal groups (such as pastoralists or ethnic tribe) may experience an absolute depravation (where they are actually separated from the resource for livelihoods) or relative depravation (where groups feel they are entitled to a resource and are denied access) from the resource.[14] Through a snowball effect, the denial of groups to important natural resources will lead to economic and political marginalization often cited reasons for conflict[15] However, a developed state with well-established social adaptation institutions is capable to alleviate the consequences of resource scarcity and it is less likely have internal conflicts erupt over scarce resources. Future studies, therefore, should focus on poor states with a history of internal violence.

---

13. Raleigh, C. and H. Urdal. "Climate change, environmental degradation and armed conflict." Political Geography Magazine 2007, 26(6): 674-694.
14. Colin H. Kahl, States, Scarcity, and Civil Strife in the Developing World, Princeton University Press 2008), p. 99.
15. Homer-Dixon, environment, scarcity and violence, (Princeton University Press 2001), p.88.

Additionally, researching connections between globalization and environmental degradation will be vital to understand why certain areas of the world are being changed faster than others.

In the international policy realm, climate change has gained more attention in the past several years. With the release of the Fourth Assessment Report of the Intergovernmental Panel on Climate Change, the scientific community has verified that the warming of the climate system is unequivocal, as is now evident from observations of increases in global average air and ocean temperatures, widespread melting of snow and ice, and rising global average sea level.[16] Additionally, in 2007, the publishing of the Stern Review found that the benefits of strong early action considerably outweigh the costs[17]. In particular, both reports strongly agree that developing countries will be the hardest hit by climate change due to lack of financial, technical and institutional capacity. Importantly, the Stern review notes: "Climate-related shocks have sparked violent conflict in the past, and conflict is a serious risk in areas such as West Africa, the Nile Basin, and Central Asia"[18]

Analyzing the linkages between climate change and conflict has found a renewed interest. Despite the renewed interest in the climate change and security nexus, the phenomenon is not new. In the 1970s, both Richard Falk's and Lester Brown's explored the connections between security and the impending climate change.[19] Specifically, Falk pointed out the relationship between time and climate change: the faster the rate of change, the less time to adapt[20]. As a result, without proper institutional capacity to manage environmental changes, the risk for violent conflict increases especially in weak states.

---

16. Ibid, p. 10.
17. Raleigh, C. and H. Urdal (2007). "Climate change, environmental degradation and armed conflict." Political Geography **26**(6): 674-694.
18. Ibid.
19. Kahl. States, Scarcity, and Civil Strife, p.108.
20. Dixon, environment, p. 121.

Another prominent scholar, Jon Barnett, explores the linkages between climate change and conflict: the political scale, the nature of governance, and the nature of environmental (as opposed to resource) changes affected by climate change[21]

Barnett stresses the importance of understanding that nation-states are unlikely to declare war with one another due to climate change effects; however, it is most likely conflicts will erupt within the state between different stakeholders[22]. Barnett mentioned that conflicts in which environmental change appears to be a contributing factor tend to be within state rather than between states[23].

Another school of thought is the Zurich group around Bächler and Spillman. This group of scholars presented a final report in 1996 based on qualitative case studies on developing countries that had to deal with both environmental problems and armed conflict. The basic assumption of this group is that environmental change may lead indirectly to conflict by intensifying the existing potential for socio-economic conflict to the point of violent escalation. According to this view, conflicts are primarily socially or politically motivated, not an irreversible consequence of environmental change. The particular aim of this study was to devise a typology of conflict that links a particular kind of environmental degradation to its socio-economic consequences and the affected parties to the conflict. Drawing on an analysis of 40 environmental conflicts, the following categories were developed:

- Regional, cross-border and demographically-induced conflicts
- Migration conflicts
- International water conflicts
- Conflicts arising from distant sources

21. John Barnett, The *Meaning of Environmental Security: Ecological Politics and Policy in the new security era*, (Zed books Ltd 2001), p.62.
22. Ibid 64.
23. Ibid 55.

The Zurich group typology shows that contextual factors other than the impacts of resource degradation ultimately determine whether competing actors will seek a peaceful or a violent solution to conflict. Among the most important socio-economic factors identified, was the lack of societal mechanisms for regulating conflict and the influence of past conflict.

A third group involved in this debate is "Oslo Group" around Gleditsch. At the International Peace Research Institute in Oslo Gleditsch started an independent quantitative research approach as a process of critical engagement with the studies of Homer-Dixon and the Zurich group. His aim was to counter the excessive complexity of the qualitative models and to provide a correction to their deficiencies regarding the selection of case studies, in particular the tendency to study countries with acute conflicts over resources (Gleditsch, 1998). According to Gleditsch compared cases in which resource conflicts are conducted violently with cases in which there is no escalation of violence. Gleditsch argues that an abundance of resources is more likely to lead to violent conflict, because rebel groups, for example, draw their funding from the exploitation of natural resources. Gleditsch emphasizes that, environmental stress is only one of several variables that may contribute to the escalation of conflict. While environmental factors such as deforestation, soil degradation and water scarcity increase the risk of violent conflict within states, economic and political factors remain crucial as explanations for the outbreak and intensity of such.

Background on Sudan

At the period of this study (before the secession of South Sudan), The Republic of Sudan was the second largest country in Africa. It's highly diverse landscape ranges from desert to tropical forest, and its abundant natural resources include oil, timber, metals and extensive agricultural land, and marine and inland fisheries. The country is also culturally diverse, as it bridges the Islamic culture of North Africa with the largely Christian south, and comprises hundreds of distinct tribal and ethnic groups. Sudan located in northeast of Africa, after being spilt it measures about 1,890,000 km with population of 31, 6 million. Its neighbors are Chad and the Central African Republic on the west, Egypt and Libya on the north, Ethiopia and Eritrea on the east, Kenya, Uganda, and Democratic Republic of the Congo on the south. The Red Sea washes about 500 mi of the eastern coast[24]. It is traversed from north to south by the Nile, all of whose great tributaries are partly or entirely within its borders. Unfortunately, Sudan has long been plagued by civil war and regional conflict. In the fifty years since achieving independence, the country as a whole has been at peace for only eleven years (1972-1983). While a historic peace agreement was reached for Southern Sudan in 2005, conflict raged on in Darfur.

Sudan has experienced several severe droughts in the past thirty years, and food production in many regions has dropped at the same time as the population has increased. The combined impacts of conflict and food insecurity have caused over five million Sudanese to be both internally and internationally displaced into camps and urban fringes, and over five million to receive international food.[25] Sudan comprises hundreds of ethnic and tribal divisions and language groups. Despite relatively abundant natural resources, Sudan is currently a very poor country due to underdevelopment, conflict and political instability.

---

24. http://news.bbc.co.uk/2/hi/middle_east/country_profiles,[accessed on 10/6/2012 at 15:11].
25. BBC (2008) "Country Profile: Sudan.

In 2004, the gross domestic product per person was estimated at 740 USD (using Purchasing Power Parity figures), as compared to 3,806 USD and 1,248 USD for neighboring Egypt and Kenya respectively. While the production and export of oil are growing significantly in importance, Sudan's primary resources are agricultural. Sorghum is the country's principal food crop, and livestock, cotton, sesame, peanuts and gum Arabic are its major agricultural exports. Sudan, however, remains a net importer of food and a major recipient of food aid. Average monthly temperatures in Sudan vary between 26°C and 36°C. The hottest areas, where temperatures regularly exceed 40°C, are found in the northern part of the country. The dominant characteristic of Sudan's climate is a very wide geographical variation in rainfall. In the north, annual precipitation ranges from close to zero near the border with Egypt, to approximately 200 mm around the capital, Khartoum. Its territory crosses over 18 degrees of latitude, which results in an extremely diverse environment ranging from arid desert in the north to tropical forests in the south. The majority of Sudan is very flat, with extensive plains in an altitude range of 300 to 600 m above sea level. Twenty-nine percent of Sudan's total area is classified as desert, 19 percent as semi-desert, 27 percent as low rainfall savannah, and 14 percent as high rainfall savannah, 10 percent as flood region (swamps and areas affected by floods) and less than one percent as true mountain vegetation.[26] Sudanese civilization dates back to at least 3000 BC. It long concentrated along the northern reaches of the Nile River, the area that came to be known as Nubia. The region's three principal kingdoms were converted to Coptic Christianity by missionaries in the 6th century AD. These Black Christian kingdoms coexisted with their Muslim Arab neighbours in Egypt for centuries, until the influx of Arab immigrants brought about their collapse in the 13th to 15th centuries. Sudan was then partly converted to Islam.

---

26. UCDP (2008) Database www.ucdp.uu.se/database Uppsala University, [accessed 28/5/2012 at 16:12].

In 1820, the Albanian-Ottoman ruler of Egypt Muhammad Ali Pasha invaded and conquered northern Sudan.

Until its gradual suppression in the 1860s, the slave trade was the most profitable undertaking in Sudan and was the focus of Egyptian interests in the country. The government encouraged economic development through state monopolies that had exported slaves, ivory, and gum Arabic.

By 1874, Egypt had conquered all of Sudan and encouraged British interference in the region. This led to the revolt of the Mahdi, who captured Khartoum in 1885. The Mahdi died in June 1885. He was followed by Abdallahi ibn Muhammad, known as the Khalifa, who began an expansion of Sudan's area into Ethiopia, but defeated by British army in 1898 in Omdurman battle. Sudan was proclaimed a condominium in 1899 under British-Egyptian administration. From 1898, the United Kingdom and Egypt administered Sudan as the Anglo-Egyptian Sudan, but northern and southern Sudan was administered as separate provinces of the condominium. In the very early 1920s, the British passed the Closed Districts Ordinances which stipulated that passports were required for travel between the two zones, and permits were required to conduct business from one zone into the other, and totally separate administrations prevailed.

In 1943, the British began preparing the north for self-government, establishing a North Sudan Advisory Council to advise on the governance of the six North Sudanese provinces: Khartoum, Kordofan, Darfur, and Eastern, Northern, and Blue Nile provinces. Then, in 1946, the British administration reversed its policy and decided to integrate north and south Sudan under one government. The South Sudanese authorities were informed at the Juba Conference of 1947 that they would in future governed by a common administrative authority with the north. From 1948, 13 delegates, nominated by the British authorities, represented the south in the Sudan Legislative Assembly. Many southerners felt betrayed by the British, because they were largely excluded from the new government. In February 1953, the United Kingdom and Egypt concluded an agreement providing for Sudanese self-government and self-determination. The transitional period toward independence began with the inauguration of the first parliament in 1954. On 18 August 1955 a revolt in the army in Torit city in Southern Sudan broke out, which although quickly suppressed, led to a low level guerrilla insurgency by former Southern rebels, and marked the beginning of the First Sudanese Civil War.[27] On 15 December 1955 the Premier of Sudan Ismail al-Azhari announced that Sudan would unilaterally declare independence in four days' time. On 19 December 1955 the Sudanese parliament, unilaterally and unanimously, declared Sudan's independence. The British and Egyptian Governments recognized the independence of Sudan on 1 January 1956. The United States was among the first foreign powers to recognize the new state. However, the Arab-led Khartoum government reneged on promises to southerners to create a federal system, which led to a mutiny by southern army officers that sparked seventeen years of civil war (1955–1972). In the early period of the war, hundreds of northern bureaucrats, teachers, and other officials, serving in the south were massacred. Dissatisfaction culminated in a second coup d'état on May 25, 1969. The coup leader, Col. Gaafar Nimeiry, became prime minister, and the new regime abolished parliament and outlawed all political parties. Disputes between Marxist and non-Marxist elements within the ruling military coalition resulted in a briefly successful coup in July 1971, led by the Sudanese Communist Party. Several days later, anti-communist military elements restored Nimeiry to power. In 1976 and Islamic group supported by Libya tried to seize power through military invasion of capital Khartoum.

---

27. Alex de Waal, War in Darfur and search for peace, (Harvard University 2007), p.75.

In 1972, the Addis Ababa Agreement led to a cessation of the north-south civil war and a degree of self-rule. This led to ten years hiatus in the civil war. In 1976, the religious wing of Umma party (Ansars) mounted a bloody but unsuccessful coup attempt. In July 1977, President Nimeiry met with Ansar leader Sadiq al-Mahdi, opening the way for reconciliation. Hundreds of political prisoners were released, and in August a general amnesty was announced for all opponents of Nimeiry's government.

In 1983 the second North South war broke out when the Sudan People's Liberation Movement/Army (SPLM/A) initiated a rebellion. The group was based in the south and some factions of the SPLM/A had a secessionist agenda whilst their leader John Garang stated that "The SPLM/A is fighting to establish a united Socialist Sudan". After that the situations was deteriorating in the country in all aspects of life which led to social unrest and then to public revolution that ousted Numaeri on April 6, 1985, then a group of military officers, led by Lieutenant General Abdar Rahman Siwar adhahab, took office for transitional period, Three days later, Adhahab authorized the creation of a fifteen-man Transitional Military Council (TMC) to rule Sudan.

After general election, Sadiq al Mahdi the leader of Umma Party formed a coalition government in June 1986 with the Democratic Unionist Party (DUP), the National Islamic Front (NIF), and four southern parties. Unfortunately, however, Sadiq proved to be a weak leader and incapable of governing Sudan. Party factionalism, corruption, personal rivalries, scandals, and political instability characterized the Sadiq regime. After less than a year in office, Sadiq al Mahdi dismissed the government because it had failed to draft a new penal code to replace the sharia, reach an agreement with the IMF, end the civil war in the south, or devise a scheme to attract remittances from Sudanese expatriates. To retain the support of the DUP and the southern political parties, Sadiq formed another ineffective coalition government[28].

In 1989, it appeared the war would end, but a coup d'état brought a military junta into power which was not interested in compromise. The leader of the junta, Omar al-Bashir, consolidated his power over the next few years, declaring himself as a president.

The fighting intensified between the government and SPLA for fourteen years till the 9th of January 2005 when the parties signed the Comprehensive Peace Agreement (CPA) in Nairobi. The comprehensive agreement opened an unprecedented window of opportunity to turn the devastation of 55 years of war, displacement, and underdevelopment into a new era of peace and prosperity. It directly addressed the key causes of the conflict, and laid out the parties' vision and commitment to accelerating development. The main provisions of the CPA include:

- The creation of a semi-autonomous region of Southern Sudan, within an overall federal structure, that also includes substantial transfers of powers from national to state level;
- Democratic elections nationwide in 2009, based on a new census and electoral roll;
- A referendum in the South in 2011, on whether the region should retain its semi-autonomous status, or progress to full independence.
- The establishment of a Government of National Unity (GoNU) for the country as a whole, and a Government of Southern Sudan (GoSS);
- A wealth sharing protocol, building on the emergence of oil as a major source of revenue;
- Two multi-donor trust funds (MDTFs), one for southern Sudan, and one for other areas of the country affected by the conflict, to finance reconstruction and development activities.

---

28. De Waal, Sudan: the Turbulent State" in de Waal (cd) War in Darfur and the Search for Peace p. 20.

Implementation of the CPA has been uneven after initial progress in setting up the Government of National Unity (GoNU), the Government of Southern Sudan (GoSS) and the national and state assemblies. In 2010 Sudan elections were held from April 11 to April 15, the elections entailed national presidential and parliamentary, the South Sudanese presidency, state governors, the Southern parliament and state assemblies. The voter registration recorded 16,336,153 voters. Over 16,000 candidates entered the contest including nearly 1,400 as independents. A total of 72 political parties registered for the election but far fewer engaged in active campaigning. On April 26, President Bashir of the NCP was officially declared the winner after Sudan's election commission announced he received 6,901,694 votes, or 68.24 percent of the total valid votes. In Southern Sudan, the current President of the Government of Southern Sudan, Salva Kiir, was the winner with 2,616,613 votes or 92.99 percent of the total valid votes.

A referendum took place in Southern Sudan from January 9 to January 15, 2011, on whether the region should remain a part of Sudan or become independent. The referendum was one of the consequences of the 2005 Naivasha Agreement between the Khartoum central government and the Sudan People's Liberation Movement (SPLM). On February 7, 2011, the referendum commission published the final results, with 98.83 percent voting in favor of independence. The predetermined date for the creation of an independent state is July 9, 2011.

Eastern Sudan has also seen rebellion after the signing of CPA. The fighting in this conflict was comparatively limited, and ended through the Eastern Sudan Peace Agreement (ESPA) signed in 2006[29].

---

29. http://www.sudantribune.com/IMG/pdf/Eastern_Sudan_Peace_Agreement.pdf.[accessed on 10 May 2012.

Background on Darfur

The boundaries of Darfur region lie between longitudes 22^0 E-27^0 E and latitudes 10^0 N-16^0 N and it covers a total area of 549 square kilometers (an area as large as France). The region is located in the remote western part of the Sudan which borders Egypt, Libya and Chad. The total population of the region is estimated at between six millions[30]. The region is composed mostly of semi-arid plains that cannot support a dense population. The one exception is the area in and around Jebal Marra. This region is extremely poorly known and documented, especially the earlier periods. Documentary history is also rather sparse, (al-Idrisi), writing in the year 1154 is the first author to offer information about the region that provides any concrete detail. The Sicilian geographer described the (Tajuwa) as pagans inhabiting the region adjacent to the Nile Valley kingdoms, who possessed two towns, the first and capital being Tajuwa and a second town lying six stages away from it called Samna, which was destroyed according to a traveler in the region, by the governor of the kingdom of Nuba. The bulk of the inhabitants were nomadic with large numbers of cattle and camels, but subject to raiding by their neighbors.[31]

Daju Period

The (Daju), inhabitants of Jebel Marra, appear to have been the dominant group in Darfur in the earliest period recorded. How long they ruled is uncertain, little being known of them. According to tradition the Daju dynasty was displaced, and Islam introduced in Darfur around the 14th century by the (Tunjur).

---

30. Al-Maqrizi in Nehemiah Levztion and J. F. P. Hopkins, eds. and trans. *Corpus of Early Arabic Sources for West African History* (Cambridge, 1981, reprinted Princeton, NJ, 2000), pp 353-54.
31. Prunier, Gérard, *Darfur: The Ambiguous Genocide*, (Cornell University Press, 2005), pp. 8-24.
32. Ibid, pp. 8-24.

The Tunjur

The first Tunjur king is said to have been (Ahmed el-Makur), who married the daughter of the last Daju monarch. Ahmed reduced many chiefs to submission, and under him the country prospered. His great-grandson, the sultan Dali, a celebrated figure in Darfur histories, was on his mother's side a Fur, and thus brought the dynasty closer to the people it ruled. Dali divided the country into provinces, and established a penal code, which, under the title of (Kitab Dali) or Dali's Book, is still preserved, and differs in some respects from Quranic law.

Darfur Sultanate

(Suleiman Solon) reigned from 1603 to 1637, and was a great warrior and a devoted Muslim; he is considered as the founder of the (Keira) dynasty and the Darfur Sultanate. Soleiman's grandson, Ahmed Bukr (c.1682-c.1722), made Islam the religion of the state, and increased the prosperity of the country by encouraging immigration from Bornu and (Bagirmi). His rule extended east of the Nile as far as the banks of the Atbara.

Egyptian rule

The Darfurians were restive under the rule of Egypt, itself under British colonization. Various revolts were suppressed, but in 1879 the British General Gordon (then governor

Governor-general of Sudan) suggested the reinstatement of the ancient royal family. This was not done, and in 1881 (Slatin Bey) (Sir Rudolf von Slatin) was made governor of the province.

Slatin defended the province against the forces of the self-proclaimed (Muhammad Ahmad AMahdil) who were led by a Rizeigat Sheikh named Madibbo, but Slatin

Pasha was obliged to surrender 1883, and Darfur was incorporated in the Mahdi's dominions.[33] Following the overthrow of Almahdia period at Omdurman battle in 1898, the new (Anglo-Egyptian) Sudan government recognized (1899) Ali Dinar, a grandson of Mohammed-el-Fadhl, as sultan of Darfur, in exchange for an annual tribute of 500 British Pounds. Under Ali Dinar, who during the Mahdi's era had been kept a prisoner in Omdurman, Darfur enjoyed a period of peace and a de facto return to independence.

British rule

However, the British allowed Darfur *de jure* autonomy until they became convinced during the first World War that the sultanate was falling under the influence of the Ottoman Empire (present-day Turkey), invaded, and incorporated the region into Sudan in 1916.[34] Within Anglo-Egyptian Sudan, the bulk of resources were devoted toward Khartoum and Blue Nile Province, leaving the rest of the country relatively undeveloped.

National independence

After independence, Darfur became a major power base for the *Umma Party*, led by Sadiq al-Mahdi. By the 1960s, some Darfuris were beginning to question the neglect of the region by the Umma, despite their consistent political support. Disillusionment with the religious sect-based parties, Khatmiyya Sufi/Democratic Unionist Party in the East and Ansar/Umma in the West, led to a temporary rise of regionally-based parties, including the Darfur Development Front (DDF).[35]

---

33. Ibid. p. 33.
34. Ibid. pp. 42-44.

To this underdevelopment and domestic political tension was added cross-border instability with Chad. Premiere al-Mahdi allowed Frolinat (***Front de Libération Nationale du Tchad)***, the guerilla movement trying to overthrow Chadian President François Tombalbaye, to establish rear bases in Darfur in 1969. However, Frolinat factional infighting killed dozens within Darfur in 1971, leading Sudanese President Gaafar Nimeiryto expel the group. This was further complicated by the interest of new Libyan President Muammar al-Gaddafi in the Chadian conflict. Obsessed with the vision of creating a band of Sahelian nations that were both Muslim and culturally Arab, Gaddafi made an offer to Nimeiry to merge their two countries in 1971.[36]

Increasing instability

In a longer term cycle, the gradual reduction in annual precipitation, coupled with a growing population, had begun a cycle in which increased use of arable land along the southern edge of The Sahara increased the rate of desertification, which in turn increased the use of the remaining arable land. Drought from the mid-1970s to early 1980s led to massive immigration from northern Darfur and Chad into the central farming belt. In 1983 and 1984, the rains failed. The region was plunged into a horrific famine. 60-80,000 Darfuris walked across the country to Khartoum seeking food. The famine killed an-estimated 95,000 Darfuris out of a population of 3.1 million and it was clear that the deaths had been entirely preventable.[37]

The incompetence of the regime, combined with the start of the Second Sudanese Civil War in 1983, proved unbearable for the country and Nimeiry was overthrown on 5 April 1985 in public uprising. Sadiq al-Mahdi came out of exile, making a deal with Gaddafi, which he had no intention of honoring, that he would turn over Darfur to Libya if he was supplied with the funds to win the upcoming elections.[38] Beginning in August 1985, Libya began sending military/humanitarian convoys from Benghazi, including an 800-strong military force that set up base in Al-Fashir (Northern Darfur) and began arming the local Baggara tribes, whom Gaddafi considered to be his local allies. By the time that Libyan relations with the United States had worsened to the extent that US bombed Tripoli in April 1986, Libya was providing key logistical and air support to Sudanese offensives against the Sudan People's Liberation Army in the rebel South.

Meanwhile, the famine had severely upset the structure of Darfuri society. The farmers had claimed every available bit of land to farm or forage for food, closing off the traditional routes used by the herders. The herders, faced with watching their animals die of starvation in the desiccated landscape, tried to force the routes south open, attacking farmers who tried to block their path and shedding blood.[39] Darfur was awash in small arms from the various neighboring conflicts and stories spread of herders raiding farming villages for all of their animals or villagers who had armed themselves in self-defense.

---

35. Ibid. pp. 44-47.
36. Ibid. p. 139.
37. Ibid. pp. 47-52.

In 1994, Darfur was divided into three federal states within Sudan: Northern (*Shamal*), Southern (*Janub*), and Western (*Gharb*) Darfur. Northern Darfur's capital is Al Fashir; Southern Darfur's is Nyala; and West Darfur's is Geneina. The division was the idea of Ali al Haj, Minister of Federal Affairs (he is from Darfur and later wrote the black book and joint rebel groups) [40], he hoped that by dividing the Fur so they will not form a majority in any state that it would allow Islamist candidates to be elected.[41] According to Human Rights Watch, hostilities broke out in West Darfur in 1998. The 1998 clashes were relatively minor, but more than 5000 from Masalit tribe were displaced. Clashes resumed in 1999 when nomadic herdsmen again moved south earlier than usual. The 1999 clashes were bloodier, with many hundreds killed, including a number of tribal chiefs. The government brought in military forces in an attempt to quell the violence and took direct control of security.

A reconciliation conference held in 1999 agreed on compensation. In 2000, a clandestine group consisting mostly of Darfuris published the *Black Book*, a dissident manuscript detailing the domination of the north and the impoverishment of the other regions. It was widely discussed, despite attempts to censor it, and many of the writers went on to help found the rebel Justice and Equality Movement.

---

38. Flint, Julie and Alex de Waal, *Darfur: A Short History of a Long War*, (Zed Books 2006), p. 25.
39. Alex Cobham, "Causes of conflict in Sudan: Testing the Black Book", Queen Elizabeth House Working Paper Number 121, University of Oxford, January 2005, p. 17.

Beginning of scarcity resources and conflict in Darfur

Rainfall and drought in Darfur

A recent study published by the UN Environment Programme asserted a very strong link between land degradation, desertification and conflict in Darfur[42]. As an agricultural economy, Sudan has long been dependent on rainfall which corresponds to the ups and downs in the country's economic growth pattern (see Figure 1) evidence to that, the year 1994 was an exceptional good year for the country, with good rainfall and a bumper crop and a year without reports of major armed conflicts in both 1993 and 1994. This enabled Sudan's economy to prosper in years of relative stability. But little attention has been paid to the economic circumstances of this region that largely depends on erratic rainfall, making it vulnerable to the adverse consequences of climate change. It is estimated that a 50 to 200 km southward shift of the boundary between semi-desert and desert has occurred since first records were collected in the 1930s[43]. Whereas an estimated 42 percent of the total area is considered cultivatable land, only 7 percent of that portion is actually cultivated. Average rainfall in Sudan varies greatly by region, with annual rainfall of less than 25 mm in the North, while Darfur has experienced below-normal annual rainfall for several years that led to the increasing tensions over scarce water resources and pasture. Vast swamp regions in Sudan also affect the water stability in the country as they greatly increase the amount of evaporation that occurs annually (roughly three times as much water evaporates due to the swamps as is annually available for use)[44].

---

40. United Nations Environment Programme, "Sudan Post-Conflict Environmental Assessment 2007, Page 93. Available at http://postconflict.unep.ch/publications/sudan/04_conflict.pdf [accessed at 20 October 2012].
41. Environmental Degradation Triggering Tensions and Conflict in Sudan (Nairobi: UNEP, 2007), Page 8.
42. Kassa, A., 1999. Drought risk monitoring for the Sudan using NDVI. MSc Dissertation, University College, London. Available at http:/www.soas.ac.uk/Geography/Water Issues.[accessed at 25 October 2012].

Figure 1: Rainfall at four stations in Darfur (annual mm.) 1940- 2001

(Dashed lines are means for 1940-71 and 1972-2002)

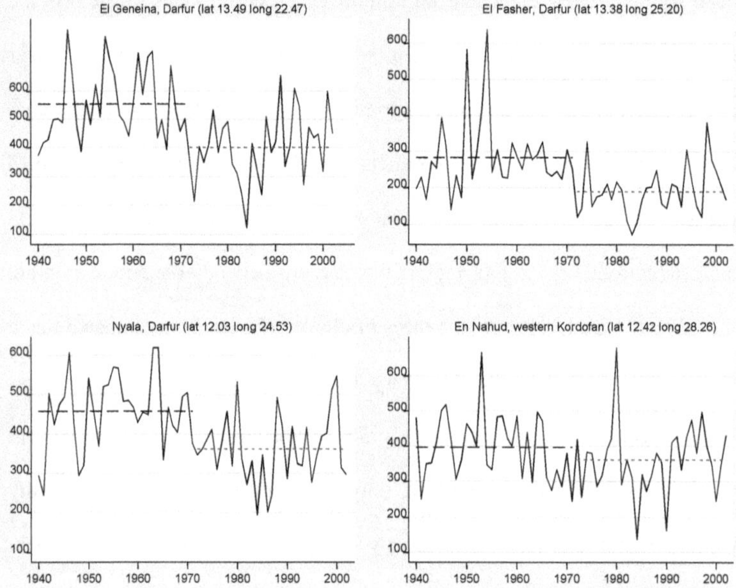

Source: Sudan rain station data provided by David Lister, Climatic Research Unit, University of East Anglia

The Inter-governmental Panel on Climate Change (IPCC) issued a report in 2001, stated that, global temperatures are predicted to increase by 1.4 - 5.8^0 C over the coming century which will, in turn, cause changes in the distribution of rainfall, the frequency and intensity of extreme weather, and a rise in sea-levels, and that would affect many human activities particularly agriculture, water sources, industry and human health[45.] Africa will be affected by the change in climate quite severely, owing to the impact of dramatic swings between drought and floods. The people of the Sahel have suffered long periods of drought during the last century: in 1903-1905, 1911-1914, 1966-1974 and 1979-1987 [46]. The long period of drought had far reaching consequences on the ecosystem of region. Rainfall is unreliable in the Sahel and the region is well known for its dual environmental problems of drought and desertification (Agnew & Chappell, 1999)[47]. Fifty percent of the perceived periods of drought in the Sudan were meteorologically confirmed with rainfall well below average [48]. In most cases famines and outbreaks of diseases came as a result.

Trends in rainfall are ambiguous. Most argue that there has been a persistent decline in most parts of the Sudan, while others argue that there has been a structural break around 1972, and that pre- and post-1972 rainfall readings oscillate around their respective means. A study of the three Darfur states confirmed that rainfall has decreased over the period for which records exist. [49]

---

43. IPCC third assessment report (2001). Climate Change Impacts, Adaptation and Vulnerability. Available at < http://www.ipcc.ch/publications_and_data/publications_and_data_reports>. [accessed at 10 September 2012].
44. Sjöström, M. (2004). Investigating vegetation changes in the African Sahel 1982-2002: a comparative analysis using Landsat, MODIS and AVHRR remote sensing data. Seminar series nr 106. Geobiosphere Science Centre and Physical Geography and Ecosystems Analysis. Lund University.
45. Agnew C.T & Chappell A. (1999). Drought in the Sahel. GeoJournal, 48, 299–311.Contribution of Working Group II to the third assessment report of IPCC. Cambridge University Press, Cambridge.
46. Teklu, T., Joachim von Braun, Zaki, E. (1991). Drought and Famine Relationships in Sudan: Policy Implications. Available at <www. citeseerx.ist.psu.edu/viewdoc/download?doi=10.1.1.78> [accessed at 11 September 2012].
47. Tearfund. (2007). Darfur: water supply in a vulnerable environment p 17. Available at < http://postconflict.unep.ch/publications/UNEP_Sudan_Darfur_Water_Resources_TF.pdf> [accessed at 12 September 2012].

UNEP (2007) found that there is a major and long-term drop in precipitation in Northern Darfur (30% over 80 years), which has led to reduced productivity.

The continuous aridity, which was marked by a decline in the water table and low vegetation cover, intensified the process of desertification. The process was further exacerbated by human responses to the changing environment as farmers continued expansion of cultivation into marginal and fragile areas to make up for declines in yields and resorted to generating income from the sale of tree crops to support their falling income[50]. Many other factors, including species composition, animal grazing, and human factors like migration, also play a significant role[51].

Climate change models differ in their prediction of the general trend of rainfall in the Sahel, but agree that variability is set to increase and that the duration of rainy periods is set to shorten. The longer term prospects are unclear, but climatic change modeling indicates that rainfall is likely to become even more variable with shorter rainy seasons in the future. This is likely to lead to an increase in frequency of failed harvests.

With regard to water availability, the IPCC (2001a: 193) stated that change "has the potential to induce conflict between different users" within the same area or between different parts of the river basin. Where there are disputes, the threat of climate change is likely to exacerbate, rather than ameliorate matters because of uncertainty about the amount of future resources that it engenders. One major impact of climate change for agreements between competing users (within a region or upstream versus downstream) is that allocating rights in absolute terms may lead to further disputes in years to come. The impact of climate change for water resources, requires more research with a focus on: a) creation

---

48. Teklu, T., Joachim von Braun, Zaki, E. (1991). Drought and Famine Relationships in Sudan: Policy Implications.
49. eaquist, J. W., Hickler, T., Eklundh, L., Ardö, J., and Heumann, B. W.: Disentangling the effects of climate and people on Sahel vegetation dynamics, Biogeosciences Discuss., 5, 3045-3067, doi:10.5194/bgd-5-3045-2008, 2008.

of credible climate change scenarios; b) characterization of natural and hydrological variability, c) improved hydrological models, d) characterization of uncertainty, e) impacts on real world water systems and f) effects of adaptation.

Effective adaptation to climate change in the water sector requires, according to the IPCC, efforts in five main areas: a) data for monitoring; b) understanding patterns of variability; c) analytical tools; d) decision tools and e) management techniques.

The water container queue at a wellpoint in Abu Shouk camp. Each water point services over a thousand people. Source: UNEP website

Land degradation in Darfur

Desertification, as defined in the UN Convention to Combat Desertification, is the degradation of land in arid, semi-arid and dry sub-humid areas caused by climatic change and human activities[52]. According to UNCCD, over 250 million people are directly affected by land degradation. In addition, some one billion people in over 100 countries are at risk. These people include many of the world's poorest, most marginalized, and politically weak citizens[53].

Long-term food productivity is threatened by soil degradation, which is now severe enough to reduce yields on approximately 16 per cent of the agricultural land, especially cropland in Africa. Sub-Saharan Africa has the highest rate of land degradation. It is estimated that losses in productivity of cropping land in sub-Saharan Africa are in the order of 0.5-1 per cent annually, suggesting productivity loss of at least 20 per cent over the last 40 years Africa is particularly threatened because the land degradation processes affect about 46 per cent of the continent.

According to UNCCD, the consequences of land degradation include undermining of food production, famine, increased social costs, decline in the quantity and quality of fresh water supplies, increased poverty and political instability, reduction in the land's resilience to natural climate variability and decreased soil productivity.

It is estimated that the livelihoods of more than 1 billion people globally are at risk from desertification, which may eventually force 135 million people off their land", according to UNEP's Africa Environment Outlook[54].

---

50. http://www.unccd.int/en/regional-access/Africa/Pages/africa.aspx [accessed on 31October 2012]
51. http://www.wmo.int/pages/themes/wmoprod/documents/WMO989E.pdf [accessed on 31 October2012]
52. UNEP, 2006. *Africa Environment Outlook 2 – Our Environment, Our Wealth.* Division of Early Warning and assessment (DEWA), United Nations Environment Programme (UNEP). Nairobi, Kenya.

In northern Sudan, there is high awareness of the issue of desertification within the academic community, and historical evidence of a number of attempts to quantify and/or limit the extent of the problem since at least the 1950s. As early as 1953 a landmark study discussed several of the sources of the problem (such as overgrazing), as well as its implications (long-term damage and reductions in productivity) [55]. UNEP considers that three compounding desertification processes are underway in Sudan, which is relatively difficult to distinguish, separate and quantify on the ground:

1. Climate-based conversion of land types from

Semi-desert to desert.

The scale and duration of the reduction in rainfall noted above is sufficient to have changed the natural environment, irrespective of human influence. This type of change occurs as a regional process, where less drought-resistant vegetation gradually dies off or fails to reproduce, resulting in a lower-density mix of different species. In a shift as rapid as that observed in Northern Darfur, this is manifest first and foremost in the widespread death of trees during drought events, which are not followed by recovery[56].

2. Degradation of existing desert environments, including *wadi*s and oases.

At least 29 percent of Sudan is already true desert. Within this large area, however, are hundreds of smaller wetter regions resulting from localized rainfall catchments, rivers and groundwater flows. Virtually all such areas inspected by UNEP were found to be moderately to severely degraded, principally due to deforestation, overgrazing and erosion[57].

53. Stebbing E.P. *The Creeping Desert in the Sudan and Elsewhere in Africa* (McCorquodal & Co 1953), p.125
54. UNEP Sudan post-conflict assessment report 2007.
55. Ibid.

3. Conversion of land types from semi-desert to desert by human action

Over-exploitation of semi-desert environments through deforestation, overgrazing and cultivation results in habitat conversion to desert, even though rainfall may still be sufficient to support semi-desert vegetation. In Sudan, a particular problem has been the conversion of dry and fragile rangelands into traditional and mechanized cropland. A detailed analysis of these processes is provided in Chapter 8. Regional differences in soil types and topography also play a part in this complex three-pronged process. The soil in the north and west of Sudan, for instance, is sandy and prone to water and wind erosion, while the south and east have more resistant clay soil. In addition, mountain ranges such as the Jebel Marra plateau form high rainfall watersheds in otherwise arid areas.

To summarize, there is sufficient disseminated evidence to support the following findings[58]:

- Moderate to severe land degradation is ongoing in the desert and semi-arid regions that cover the northern half of Sudan;
- A 50 to 200 km southward shift of the boundary between desert and semi-desert has occurred since rainfall and vegetation records began in the 1930s. This shift, however, has not been well quantified and is based largely on anecdotal evidence and small-scale studies;
- The desert and semi-desert boundaries are expected to continue to shift southwards due to declining precipitation/reliability of precipitation;
- Most of the remaining semi-arid and low rainfall savannah on sand, representing approximately 25 percent of Sudan's agricultural land, is at considerable risk of further desertification, to the extent that food production in these regions will at minimum

56. Ibid.

plateau, and more likely continue to drop significantly (i.e. up to 20 percent or more); and

• Modeled predictions of a future 70 percent drop in food production in Northern Kordofan have actually already taken place on a smaller scale and on a short-term and local basis, due to reduced rainfall and ongoing land degradation and abandonment. This trend is expected to worsen with time and the predicted result is that in the absence of major changes in agricultural patterns, food insecurity will only increase in these regions. The area at greatest risk is the Sahel belt, as shown in Figure 2.5. It includes the conflict-affected parts of Darfur, the previously drought-stricken parts of Northern Kordofan and Khartoum states, and conflict- and drought-stricken Kassala state. Much of the evidence for the above findings is piecemeal, anecdotal and/or based on site-specific data. The limited numerical data available does validate the anecdotal findings, but further solid and comprehensive analysis is clearly needed. (See figure 2)

Figure 2: Soil degradation in Darfur

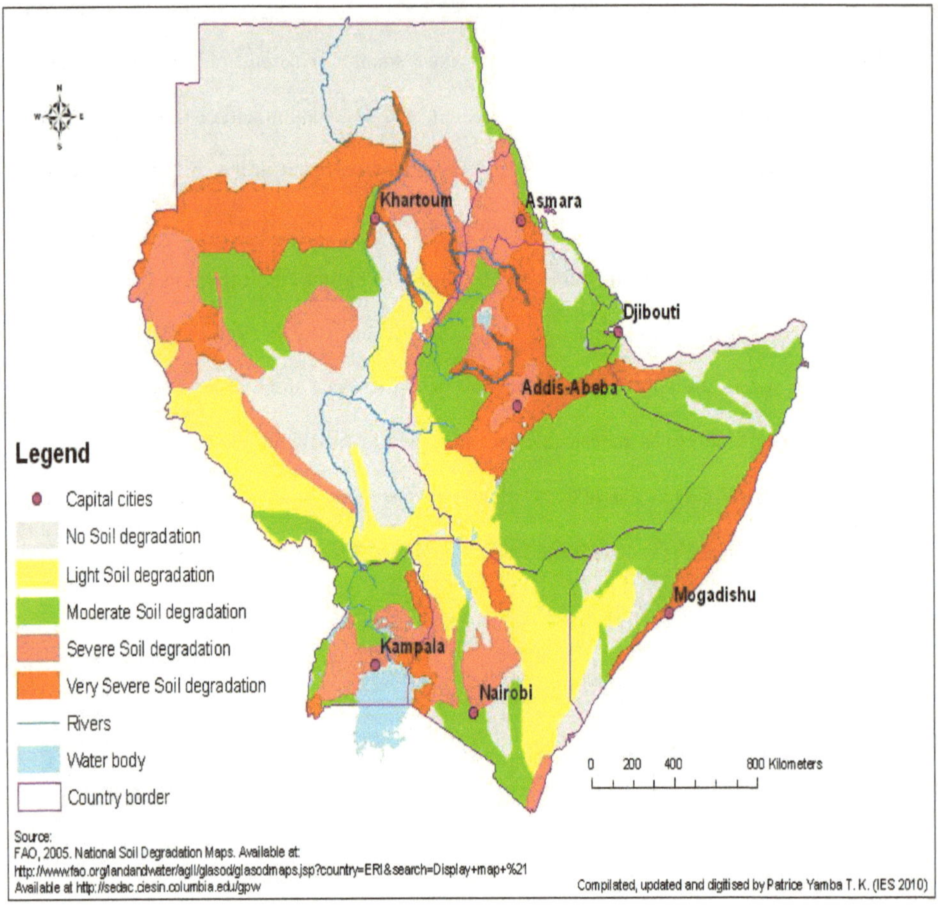

Source: FAO, 2005, National Soil Degradation Maps. Available at: <www.fao.org

Land tenure system in Darfur

Issues of land in Darfur include questions of "ownership" (*milk*), tenure and use goes back some three to four hundred years. The tensions between different tribes and between nomads and sedentary communities over grazing rights, migratory routes and interactions along and across the sedentary/nomad frontier is not a new phenomenon but the scale of conflicts on land has increased due to the effect of climate change on land productivity, water resources and land tenure change.

Range and pasture are the backbone of livestock production in the Sudan in general, and Darfur in particular. The growth of forage plants and grazing grasses undoubtedly depends on rainfall. Therefore, due to drought and the shortage of rains, the productivity of natural pasture has decreased.[59] The deterioration is not limited to grasses, but includes a decrease in the production of trees and in particular their pods and leaves, which normally contribute over 30 percent to range carrying capacities. Regarding the decrease in pasture lands, while there have been substantial decreases in the *qoz* soils, losses in the clay and *wadi* lands are double those of the *qoz* soils.

Evolution of land tenure system in Darfur

Land tenure arrangements in Darfur have evolved over centuries and are largely based on an original system in which communal families are given usufruct rights to farm land based on their needs, providing it is regularly cultivated. According to this system, the community leader (normally the village head-man) is responsible for land allocation and for recognizing new occupants.

During the Keira dynasty of Sultan Musa Ibn Suliman (1680-1700), a new system of granting land titles called *hakura* was introduced. Administrative *hawakeer* were usually

---

57. Fadul, A. A. *Natural Resources Management for Sustainable Peace in Darfur. Sudan, University for Peace, (2004)*, p. 34.

granted to tribal leaders and came to be known as *dar*. This confirmed communal ownership of land for a given group of people who usually made up a tribe or sub-tribe under a recognized leader. The *hakura* system effectively managed the allocation of land among tribes and regulated access and use of natural resources in a harmonious way during times of plenty and maintained a complementary economy based on the exchange of water and access to pasture land for meat and dairy. However, in times of scarcity, the system came under great stress. Increased demand for access to grazing land and water from nomadic tribes caused a lot of skirmishes between herders and farmers.

To distinguish the territories in the absence of charters, tribal homelands names usually attributed to the tribe which traditionally owns the specific piece of land, for instance "Dar Zaghawa" to mean land of the Zaghawa people and Dar Rizaygat to mean land of the Rizaygat people. In addition to identifying land ownership, this branding of territory based on tribal affiliation meant that over time, the land became a symbol of group identity in addition to being an economic asset.[60]

During the Turkish colonialism (1821-1884) and the brief Mahdist rule (1885-1898), no radical changes were introduced to the basic structure of the land tenure system. Nevertheless, the Mahdist government transferred the ownership of vast tracts of land from disloyal to loyal groups.[61]

Wars, droughts and famines during the Mahdist rule had resulted in tribal migrations, abandonment of cultivation, arrogation of rights and a general decline in agriculture. Following the re-conquest, the colonial authorities proclaimed that no land sales would be recognized without a valid title; thereby it forbade the selling of land without the written

58. O'Fahey, R.S. and Abu Salim M.I. *Land in Dar Fur: Charters and Related Documents from the Dar Fur Sultanate.* London: Cambridge University Press *1983.*
59. Awad, M. H. (1971). The evolution of landownership in the Sudan. *The Middle East Journal*, 25(4), 212-228.

consent of the provincial governors, and declared that all sales for which such consent was not given would be regarded as null and void. [62]

The colonial government promulgated an ordinance for the settlement of rights over forest and unoccupied lands. According to that ordinance, the government was deemed the legal owner of all land and forests which were entirely free from private rights or that the rights existing, do not amount to full ownership.

Substantial change came in 1971 under the Nimeyri regime, with the abolition of Native Administration and, importantly, of the chiefs and their courts. President al- Nimeyri regarded both to be retrograde and un-modern and sought their replacement by young graduates from the University of Khartoum's School of Public Administration alongside a more "modern" town-council style of administration in urban areas.[63]

Climate change casual relation with land tenure

In order to understand the impact of ecological change in Darfur on the current state of war in the region, one must look at the societal components of Darfur and their livelihood pattern. The Darfurian community consists of settled farmers and the pastoralist nomads; throughout the history they maintained solidarity/strife relationship with each other. They exercise mutual solidarity in times of normal hardship, but in times of severe hardship, when bodily survival is literally at stake, they engage in mortal combat.[64]

---

60. The foregoing account was based on Mr. B.H. Bell's "Note on tribal lands", August 10, 1930, enclosed with a letter from D. Newbold, Kordofan Governor, addressed to all District Commissioners, May 28, 1933; El Obied (Kn.P./1/f.1/34).
61. Rünger, M. *Land Law & Land Use Control in Western Sudan: The case of southern Darfur*. London: Ithaca Press 1987).p.5.
62. Suliman, M. 1998, The Inversion of Ethnicity from Perception to Cause of Violent Conflict, Friedensbericht, Ruegger Verlag, Zurich 1998), p.10.

This relatively calm setting of Darfur was strongly disrupted during the 1980s by the prolonged drought which has persisted with only minor interruptions since 1967 - and the ensuing and unprecedented mass population movement, impoverishment and destitution of the inhabitants of the affected arid and semi-arid zones.[65]

Darfur has witnessed five drought disasters over the last hundred years. Two of these, however, have occurred in the last twenty years of the 20th century alone. In this dry region lying between the isohyets 100mm and 600mm - a mere 100mm decline in the mean annual precipitation is quite sufficient to bring people and livestock to the brink of disaster. Rainfall data covering the period 1950-1990 reveal three major spans of drought, a relatively mild one in the mid-1960s, and a slightly sever one in 1973-1975 and a very severe one in 1982-1984. In all three cases the draught was accompanied by flaring of skirmishes, the worst of which took place in the mid-1980s.[66] The diagram below (figure 3 M.suliman) correlates rainfall data to conflict intensity over a 30 year period (1957-1987). The diagram reveals two interesting patterns: an increase in incidents of conflict with the corresponding decrease in rainfall and a lag between minimum rainfall and maximum conflict intensity of roughly one year, a relaxation period for the impact of the draught to take full effect (see figure 3 below).

---

63. ElNur, I. (et al) (1992) Displaced and Refugee Studies in the Sudan an Annotated Bibliography, University of Juba 1992), p.89.
64. Suliman, M. The conflict in Darfur: resource dimension. Available at <www.sudan-forall.org/.../Dr-Mohamed-Suliman.pdf> [accessed on 17 July 2012].

Figure 3: Rainfall and conflict correlation

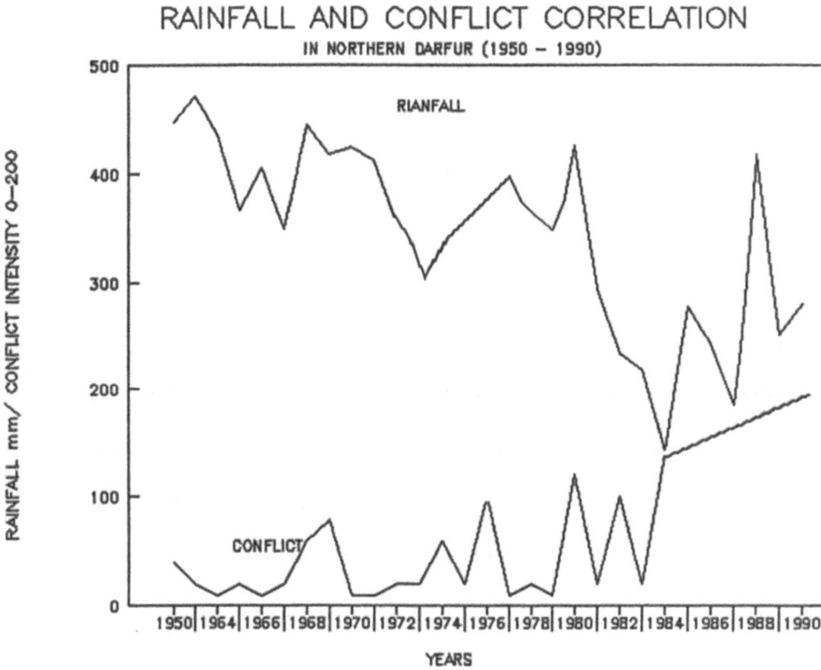

Suliman, M. The conflict in Darfur: resource dimension. Available at <www.sudan-forall.org/.../Dr-Mohamed-Suliman.pdf> [accessed on 17 July 2012].

The diagram also exposes an anomaly between the impact of the draught of the mid-1970s and that of the mid-1980s, which were of medium and severe intensity respectively, but the latter caused far greater social turbulence (see figure 1). The drought of the 1980s brought famine, displacement and armed conflict on a much larger scale than that of the 1970s. Possible explanations of this apparent discrepancy could be:

I) In the 1970s, the agricultural food production of the Sudan was geared towards the internal market, while in the 1980s food production was geared towards export.

II) During the 1970s, the regional food and other reserves of Darfur mitigated the impact of the drought. The 1980s found these reserves depleted.

III) In the 1970s, the local traditional administration was still functioning and supportive. In the 1980s, it was abolished by the government of Numerei.

IV) In the 1970s, there was no large-scale warfare in the Sudan or neighbouring countries, while the 1980s saw the emergence of the civil war in the Sudan and Chadian-Libyan war had been large-scale and wide-spread. With the onset of the drought, rural economies begin to fall apart. Livestock began to die in large numbers and livestock owners were forced to dispose of their remaining herds for very little. City merchants turned away from the collapsing rural economy, leaving it to its own fate. Abandoned by both nature and the market, famine and poverty took hold. Life became a real struggle. At this point, rural society was ripe for dislocation, turbulence, and, ultimately war. That is exactly what happened in northern Darfur in the mid-1980s.

What danger does climate change pose to societies?

Migration

Professor Myers' estimate of 200 million climate migrants by 2050 has become the accepted figure cited in respected publications from the IPCC to the Stern Review on the Economics of Climate Change.[67] This is a daunting figure; representing a ten-fold increase over today's entire documented refugee and internally displaced populations.[68] To put the number in perspective it would mean that by 2050 one in every 45 people in the world will have been displaced by climate change. It would also exceed the current global migrant population. According to the International Organization for Migration (IOM) about 192 million people, or 3 per cent of the world's population, now live outside their place of birth.[69] According to the Millennium Ecosystem Assessment the total population of the world's dry lands amounts to roughly two billion. Among those, the livelihoods that need to be considered as severely affected by dry land degradation amount to at least 250 million and probably up to 1.2 billion people, depending on the choice of criteria. Many more are threatened by the prospect of desertification. In particular, they face losses in biological and economic productivity of land as a result of gradual soil degradation. Given the intricate linkages of soil degradation, agricultural production, food security and poverty, many household members may as a consequence of desertification leave their homes and seek to supplement the income of their families through cash remittances. Others may even be uprooted with their whole families and decide to migrate in order to survive.

---

65. Stern, N., (Ed.), *the Economics of Climate Change: The Stern Review,* Cambridge University Press, Cambridge, 2006, p. 3.
66. Dupont, Alan & Pearman, Graeme, 2006, "Heating up the planet: Climate change and security", Lowry Institute for International Policy, Paper 12, Sydney, p. 55.
67. International Organization for Migration. Available at < http://www.iom.int/jahia/page3.html, [accessed 8 July 2012].

In fact, it is one of the key findings of the Millennium Assessment that recurring droughts and land degradation are predominant factors in the movement of people from dry lands to other areas. Rather unsurprisingly.

Desertification + Migration= Conflict

The nexus between desertification and migration is also considered as a security issue. However, growing demands on limited resources will bring about more conflicts of interest. Fierce competition for scarce resources of land and water is not only a cause for migration, but can also be intensified if incoming migrants or refugees increase the pressure on fragile environments and rural communities that are already under stress. Although so far no straight forward link between migration and conflict has been proven, however, when this migration occurs in such region already affected by climate change, resources scarcity and complexity of land ownership, it's likely to find casual relation between migration and conflicts.

Degradation and population movement

Darfur has been affected by drought and desertification since the 1967. These crises caused a major societal movement as the herders live in the north migrate south, where there is better water and pasture for their animals on which their livelihood heavily depends, but the indigenous tribes who are living in these less dry places were not happy with the new comers which led to clashes between the farmers and the nomads. At this point, the concept of land ownership must be considered since every tribe has its own land Dar or Hakura and this system dates back to the 17th Century, the Sultanate of Darfur introduced a system of granting land titles (estates), called *hakura* [70], as a result, land in

68. Abdul-Jalil, M. A. (2006). The Dynamics of Customary Land Tenure and Natural Resource Management in Darfur. Available at <www4.fao.org/cgi, [accessed on 18 July 2012].

Darfur became tribal property. Sometimes, tribes move in from outside the region or across the international borders[71], due to the intertribal relations. The deterioration of natural resources coupled with poor development projects and services in addition to the spread of unemployment and Proliferation of small arms, all led to the emergence of the phenomenon of armed robbery, as a means for survival. The ensuing instability forced many tribes to relocate to other areas. However, others resorted to organizing tribal militia to protect their lives and properties.[72]

69. Dyab, Ahmed I. *Identity and Peace Issues in Sudan*. Khartoum. Dar Azza Press 2007), p. 94.

70. Fadul, Abduljabbar Abdalla (2006) '*Natural resources management for sustainable peace in Darfur.*' University for Peace: Addis Ababa.

Displaced populations settle on the outskirts of existing towns, as seen here in El Fasher, Northern Darfur, where the new settlement is distinguished by white plastic sheeting. These new arrivals add to the environmental burden on the surrounding desert environment
Source UNEP.

History of tribal conflicts over resources in Darfur

Tribal and small-scale conflicts fought only with small arms have occurred continuously throughout the history of Sudan. No part of the country has been exempt from such clashes, but they have been concentrated in the south, west and east of the country for the last thirty years. Their causes are generally poorly recorded, but include disputes over cattle theft, access to water and grazing, and local politics[73].

In 1873/74 there was a major famine called Karo Fata (white bone). The years that followed coincided with a succession of famine caused by drought conditions, poor harvests, and the frontier conflicts between Darfur and Weddai resulting from pillaging and trading. The famine of 1888-92 was possibly the worst ever – the major cause of which was fighting between large numbers of the Mahdist forces under the governor of Darfur and a religiously inspired rebel army. This was the time of the Khalifa's forced migration to Omdurman. These years resulted in impoverishment and destruction.The third region wide 'famine that killed' and the only one whose major cause was a drought occurred during 1913/14[74]. Famine was exacerbated in the south eastern Darfur by fighting between the Rizeigat and the forces of King Ali Dinar. In a letter quoted by Alex de Waal the Nazir of the Rizeigat wrote "there is not a single house that is not burned down by [the Fur armies] the famine contributed to instability; breakdown of local government, raiding for food, and enormous famine migrations.

Many studies found substantial evidence of a strong link between the recent occurrences of local conflicts in Darfur and the environmental degradation. UNEP has classified the hundreds of distinct rural social units present in the current and historical conflict regions

---

71. Wadi, A.I. *Perspectives on Tribal Conflicts in the Sudan*. (University of Khartoum 1998), p. 78.
72. De Waal A. Famine that Kills: Darfur, Sudan, 1984-85, (Oxford University press 2005), p.191.

Into two major groups, based on livelihood strategies:

1. Predominantly sedentary crop-rearing Societies/tribes;

2. Predominantly nomadic (transhumant) livestock-rearing societies/tribes;

Most of the recorded local conflicts are within and between the first two groups: pastoralists and agriculturalists fighting over access to land and water. The report stated that there is an extensive history of local clashes associated with this issue.[75] A 2003 study on the causes of conflict in Darfur from 1930 to 2000, for example, indicates that competition for pastoral land and water has been a driving force behind the majority of local confrontations for the last 70 years (see figure 4) causes of local conflicts in Darfur) From the mid-1980s Darfur witnessed a gradual increase of violent interaction between various groups. Some of these conflicts took place between nomads and nomads and others between sedentary but the most vicious has been that involving the largest sedentary groups against the largest nomadic groups.[76] This has put an end to a pattern of complementary interaction and peaceful coexistence that characterized the relationship between the two sides for decades. Access to land and natural resources has been directly associated with the majority of violent confrontations between various ethnic groups in Darfur so far. The following table gives a rough indication of the issues and groups involved in violent confrontations with each other from 1932 to 2000, just before the outbreak of the current war (in 2003). The table indicates very clearly that, camel herders of North Rizeigat, Zaghawa and Ziyadiya are the most involved in violent conflicts.[77]

---

73. Braukamper, Ulrich (2000). '*Management of Conflicts over Pastures and Fields among the Baggara Arabs of the Sudan Belt*' in *Nomadic Peoples journal,* Volume 4 issue 1.
74. Babiker, Mustafa. 2001. *Resource Competition and Conflict: Herder/Farmer or Pastoral-ism/Agriculture*? In African Pastoralism: Conflict, Institutions and Government, eds. M. A. Mohamed Salih, Ton Dietz and Abdel-Ghaffar Mohamed Ahmed, 135–144. London: Pluto Press.
75. Suliman, M. The conflict in Darfur: resource dimension. Available at <www.sudan-forall.org/>

Figure 4: History of Tribal conflicts in Darfur

Major conflicts reported in Darfur, 1932–2000

No.	Tribal groups involved	Year	Major cause of conflict
1	Kababish, Kawahla, Berti and Meidob	1932	Grazing and water rights
2	Kababish, Meidob and Ziyadiya	1957	Grazing and water rights
3	Rizeigat and Maalia	1968	Local politics of administration
4	Rizeigat and Dinka	1975	Grazing and water rights
5	Beni Halba and Mahariya	1976	Grazing and water rights
6	N Rizeigat (*abbala*) and Dajo	1976	Grazing and water rights
7	N Rizeigat (*abbala*) and Bargo	1978	Grazing and water rights
8	N Rizeigat and Gimir	1978	Grazing and water rights
9	N Rizeigat and Fur	1980	Grazing and water rights
10	N Rizeigat and Bargo	1980	Grazing and water rights
11	Ta'aisha and Salamat	1980	Local politics of administration
12	Kababish, Berti and Ziyadiya	1981	Grazing and water rights
13	Rizeigat and Dinka	1981	Grazing and water rights
14	N Rizeigat and Beni Halba	1982	Grazing and water rights
15	Kababish, Kawahla, Berti and Meidob	1982	Grazing and water rights
16	Rizeigat and Messeiriya	1983	Grazing and water rights
17	Kababish, Berti and Meidob	1984	Grazing and water rights
18	Rizeigat and Messeiriya	1984	Grazing and water rights
19	Gimir and Fellata (Fulani)	1987	Administrative boundaries
20	Kababish, Kawahla, Berti and Meidob	1987	Grazing and water rights
21	Fur and Zaghawa	1989	Armed robberies
22	Arab and Fur	1989	Grazing rights
23	Zaghawa and Gimir	1990	Administrative boundaries
24	Zaghawa and Gimir	1990	Administrative boundaries
25	Ta'aisha and Gimir	1990	Land rights
26	Bargo and Rizeigat	1990	Grazing and water rights
27	Zaghawa and Maalia	1991	Land rights
28	Zaghawa and Marareit	1991	Grazing and water rights
29	Zaghawa and Beni Hussein	1991	Grazing and water rights

30	Zaghawa v. Mima and Birgid	1991	Grazing and water rights
31	Zaghawa and Birgid	1991	Grazing and water rights
32	Zaghawa and Birgid	1991	Grazing and water rights
33	Fur and Tarjam	1991	Land rights
34	Zaghawa and Arab	1994	Grazing and water rights
35	Zaghawa (Sudan) v. Zaghawa (Chad)	1994	Tribal politics
36	Masalit and Arabs	1996	Grazing, administration
37	Zaghawa and Rizeigat	1997	Local politics
38	Kababish Arabs and Meidob	1997	Grazing and water rights
39	Masalit and Arabs	1996	Grazing, administration
40	Zaghawa and Gimir	1999	Grazing, administration
41	Fur and Arabs	2000	Grazing, politics, armed robberies

Source: Richard Rottenburg (Hg.): Nomadic-sedentary relations and failing state institutions in Darfur and Kordofan (Sudan).

As the natural resources highly became scarce, and the flow of immigrants tribes southward has enormously increased, different tribes literally applied their traditional rights as the owner of the land to block of livestock migration routes. Blocking of *marahil* (animal migration routes) became more frequent. Some researchers have noticed that nomads often complain about such a practice which is against customary land tenure arrangements.[78] Blocking of routes has become a permanent item in the agenda of tribal reconciliation conferences convened for the last two decades to solve inter-ethnic disputes in Darfur. It is one of the common causes of resource based grass-root conflicts.

Another factor that accelerates the tribal conflicts is the spread of small arms. The last two decades witnessed a huge increase in the number of small arms in the hands of civilians in Darfur despite the fact that, no statistical estimates are available.[79]

Supplies flow from army stores (corruptive practices) and neighboring countries (mainly Libya and Chad). The availability of arms does not in itself represent a conflict factor but rather a catalyst which in the presence of hostilities contributes to rapid escalation of violent confrontations. Small arms help spread armed robbery in Darfur which lead to inter-ethnic violence. One of the most important factors of conflict in Darfur relates to the fact that the region borders two neighboring countries (Libya and Chad) that have either been at war with each other or supported insurgent groups working across its borders. Since the 1960s Chad has constantly experienced various episodes of its long-lasting civil war. Most of the actors involved in the Chadian civil war share common ethnic identity with groups existing in Darfur.

---

76. *Land tenure and inter-ethnic conflict in Darfur*. Paper presented at the Conference on Land Tenure and Conflict in Africa: Prevention, Mitigation and Reconstruction at the African Centre for Technology Studies (ACTS), December 9–10, 2004 in Nairobi, Kenya.
77. Fadul, Abduljabbar A. 2004. *Natural Resources Management for Sustainable Peace in Darfur*. Paper presented at a Workshop organized by U. N. University for Peace & University of Khartoum Peace Research Centre on Environmental Degradation and Conflict in Darfur, December 15–16, in Khartoum, Sudan.

The nexus between climate change and the conflict in Darfur

As mentioned at the introduction, this study only focus on the impacts of climate change and the stress it puts on various societies, particularly in so-called weak and less developed states, with special focus on renewable natural resources like water and land as a key determent in intrastate conflicts within weak states. And that is because weak states lack the capacity to adapt with the impacts of disasters. Most of the scholars who tackled the relation between climate change and security, have generalized their findings on all countries while countries are incomparable and differ in many aspects. For example, the February 2010 earthquake in Chile was stronger than the one that devastated Haiti just a month before. Yet, Haiti faced more casualties and damages than Chile, because of its poor and unprotected infrastructure, lack of preparation to face such disaster, and absence of emergency response to the post-earthquake situation.[80] On the contrary, Chile maintained its building codes and built its own emergency response system for such situations.[81] Chile is a wealthier nation with the experience of tackling such seismic catastrophes.

The direct link between climate change and security has been increasingly recognized. In 2007, for example the German Advisory Council on Global Change published a report entitled "World in Transition: Climate Change as a Security Risk". Drawing on the alarming figures published by the Intergovernmental Panel on Climate Change (IPCC) in the same year ,it argued that without resolute counteraction, climate Change would significantly increase the stress put on various societies, particularly in so-called weak and fragile states

---

78. Frank Bajak, *Chile-Haiti Earthquake Comparison: Chile Was More Prepared. The Huffington Post*, February 27, 2010, http://www.huffingtonpost.com [accessed on 20 July 2012].
79. Ibid. Bajak.

UNEP's 2007 analysis indicates that there is a very strong link between land degradation, desertification and conflict in Darfur. where exponential population growth and related environmental stress have created the conditions for conflicts to be triggered and sustained by political, tribal or ethnic differences can be considered a tragic example of the social breakdown that can result from ecological collapse. Long-term peace in the region will not be possible unless these underlying and closely linked environmental and livelihood issues are resolved.[82] Moreover, climate change receives considerable attention in strategic documents of the defense forces in many countries such as the United States, Germany, France, China, Finland and Australia.[83]

Darfur faced relentless desertification over the past several decades. The process of desertification had eroded the surface clay and soils, and finally depleted the productivity of arable lands in the greater region of Darfur.[84] This environmental degradation caused forced ecological migration towards the southern part of Sudan. This internal displacement caused tensions in the issues of land use and resource sharing, which finally continued to threaten peaceful coexistence and the social cohesion of Darfur. Hence, the situation ignited local tensions and provoked violent resource-based conflicts since February 2003.

What happened in Darfur is continuous environmental calamities decrease the economic growth of the people, hampers its social cohesion, and destabilizes its political structure[85] and consequently, reduces economic opportunities for the region by causing demographic

---

80. UNEP: post conflict environmental assessment report, June 2007.
81. Michael Brzoska. Climate change as a driver of security policy, *Paper prepare for SGIR 2010*, (Stockholm: SGIR, September, 2010), pp 8-10.
82. University for Peace, *Environmental Degradation as a Cause of Conflict in Darfur (conference Proceeding)*, (Ethiopia: University for Peace, 2004), pp 12-13.
83. Shaukat Hassan. Environmental Issues and Security in South Asia,‖ *ADELPHI Paper*, no. 262, (Autumn 1991), pp 5-6.

displacement within the traditional tribal borders. The unprecedented movement of population southward in pursuit of renewable natural resources across the tribal boarders raises tension between the different tribes in Darfur. The environmental stress in Darfur and the lack of preventive mechanism increased political disorder, civil strife, and even insurgency.

The accumulation of number of environment-related factors such as environmental degradation, depletion, and lack of access to natural resources led to the outbreak of violent conflict. Environmental conflicts in this dry region was inevitable because of overuse of renewable resources, overstrain of the environment's sink capacity, and impoverishment of the space of living. Due to the long term drop in rain fall averages, resource scarcity generates more demands for the natural resources in the presence of lack of supply in response to an increasing demand and that was quite sufficient to bring primitive society to conflict in the absence of traditional rules of environmental management that have been weakened. While this scarcity and have required a high level of community management, given that different groups use resources in different ways for their livelihoods.

Beside the renewal resources scarcity, Darfur has experienced significant growth in population over recent decades, from just over 1m people in the mid-1950s to around 6.5m in the early 2000s (see Figure 5 below).

Table 1: Population in Darfur

Year	Population
1956	1,080,000
1973	1,340,000
1983	3,500,000
1993	5,600,000
2003	6,480,000

Source: University of Peace, *Environmental Degradation as a Cause of Conflict in Darfur*, December 2004, p. 35.

The marked increase in population density since the mid-1970s has put pressure on both sedentary and pastoralist livelihood systems. The UN University of Peace conference Environmental Degradation as a Cause of Conflict in Darfur', held in Khartoum in December 2004, describes the following links between the environment and conflict:

1. The increase in population density intensifies cropping and grazing.
2. This means shorter fallow periods for fields and overgrazed rangeland.
3. These processes cause deterioration in yields and carrying capacities.
4. Larger areas are needed to support the same yields and herds, but demands and herds are increasing.
5. Herders and farmers compete for access to resources, leading to conflict.

The abolition of Native Administration system since 1971 preceded the 1972-73 drought, which meant that there was reduced governance capacity to address the problems of migration and conflict that occurred at the time. The most severe drought, in 1984, triggered major migrations and changes in livelihoods, again in the absence of tribal governance ahead of the reconstitution of the Native Administration in 1986. At the same time the number of weapons in Darfur was beginning to increase and regional ethnic tension was growing. This led to one of Darfur's most significant tribal wars between Fur and Arab tribes from 1987 to 1989. This pattern of environmental conflict coinciding along tribal divides is reflected in the current conflict. This war was not an isolated event: UNEP's Post Conflict Environmental Assessment lists conflicts in Darfur since 1975 in which environment and livelihoods have been a component. Darfur cannot be reduced to a discussion over resource scarcity leading to conflict alone, but this scarcity, which is a major driver in conflict at the tribal level, must be addressed in the context of poor governance and underdevelopment.

The role of multilateralism in solving the crises in Darfur

Since the outbreak of the conflict in Darfur in 2003, the conflict has been most prominently on the international agenda and the international community has been greatly involved, using different approaches from international criminal investigations, peacekeeping and mediation. Here a question must be asked, why despite this level of attention and involvement so little progress has been made over the last nine years and whether it is possible that ill-informed outsiders with the threat of military power on their side can make things worse rather than better? It tends to be that the misalignment of strategies and objectives by the international community itself has contributed to creating the main obstacles to a settlement of the conflict: the ICC arrest warrant for President Al-Bashir and the fragmentation of the armed opposition groups.

In 3 February 1997 the Special Rapporteur on the Situation of Human Rights in Sudan, Gaspar Biro, warned of serious intertribal clashes in West Darfur between the Rizeighat, Missirya and other Arab nomadic tribes, on the one hand, and the Zaghawa and Masalit tribes on the other.

The armed conflict started on February 2003, In April 2003, two loosely connected movements, the Sudan Liberation Movement SLA and the Justice and Equality Movement JEM, started attacking the government forces in Darfur. Some of the major rebel attacks were on military bases and airports in Darfur.

Since March 2004 UN gave an attention to Darfur, the UN coordinator Mukesh Kapila called Darfur *"the world's worst humanitarian crisis"* and compared it to Rwanda. On 5 October 2005 the Special Advisor to the UN Secretary-General on the Prevention of Genocide, Juan Mendez, warned of the alarming security situation in Darfur, as well as the lack of accountability. He recommended that the Security Council and the AU elaborate a strategy to neutralize the Janjaweed in the absence of Government of Sudan cooperation.

United Nations response

The U.N. Security Council (UNSC) has adopted nine major and overall 26 resolutions since the start of the crisis and repeatedly warned and urged the Sudanese government to stop the violence. The first resolution adopted by the UNSC was 1556 of 30 July 2004 which mandated the government of Sudan to disarm the Janjaweed militias and bring them to justice. The resolution also imposed arms embargo on non-governmental entities. The UNSC then failed to monitor the implementation of its demand, let alone take action against Khartoum for failing to act. (In reality, the Sudan government was incapable of disarming the militia.)[86]. The UNSC then passed a second resolution, 1564 of September 2004 in which it urged the government of Sudan to comply with resolution 1556 or it would use Chapter VII of the UN Charter. Also, the resolution created an International Commission of Inquiry to report on violations of human rights and humanitarian law in Darfur. In 24 March 2005, the UNSC passed a third resolution, 1590, by which it established UN Mission in Sudan (UNMIS). Again, in 29March 2005, the UNSC passed Resolution 1591 which called for a ban on offensive military flights over Darfur, extended arms embargo to all in Darfur, and authorized targeted sanctions (travel ban and asset freeze). In order to exhaust all means to solve the problem of Darfur, the UNSC passed Resolution 1593 in 31 March 2005 and referred the matter to the International Criminal Court (ICC). The investigation by the ICC was stonewalled by the Khartoum.

---

84. Dewal Alex, International Affairs journal vol 83, 2007 Journal Publishing Ltd/The Royal Institute of International Affairs.

In 26 April 2006, the UNSC adopted Resolution 1672 in which it named four Sudanese officials for sanctions (travel ban and asset freeze). In 16 May 2006, the UNSC passed Resolution 1679 that called on non-signatories to sign the Darfur peace Agreement (DPA) that called for acceleration of transition to UN Operations in Darfur. In 31 August 2006, the UNSC passed Resolution 1706 that mandated the U.N. Mission in Sudan (UNMIS) to take over the African Mission in Sudan (AMIS). The resolution was rejected by the Sudan. On 27 February the Chief Prosecutor of the International Criminal Court (ICC) presented evidence against the first suspects in his investigation of mass atrocities in Darfur. 7 June: Luis Moreno Ocampo, Prosecutor of the International Criminal Court, urged the Security Council to take the lead in calling on the Khartoum Government to arrest Humanitarian Affairs Minister Ahmed Haroun and Janjaweed militia leader Ali Muhammad Ali Abd al-Rahman, also known as Ali Kushayb. They are wanted on 51 counts of war crimes. In 31 July 2007, the UNSC unanimously passed Resolution 1769 for the establishment of the joint A.U.-U.N. Hybrid Mission in Darfur (UNAMID). On July 31, 2008, the UNSC passed Resolution 1828 in which it extended the mandate of the UNAMID for a further 12 months till July31, 2009. On July 30, 2009, the UNSC adopted Resolution 1881 extending the mandate of the UNAMID for one more year.

In 4 March 2009, ICC issued arrest warrants for the President of Sudan, Omar al-Bashir, accusing him of war crimes and crimes against humanity. Khartoum refused to accept the ICC decision and stated that since it has not ratified the Rome Statute of the ICC. Another setback the ICC decision has received is the response of the African Union. The African Union stated in its 13th summit report that it was ending cooperation with the ICC for charging President Omar al-Bashir with war crimes. The AU added that it deeply regretted that the "UN ignored its previous demand for the ICC in the Hague to postpone its arrest warrant against President al-Bashir for crimes in Darfur." Moreover, the ICC decision has resulted in the expulsion and harsh treatment of international NGOs. A growing number of humanitarian organization and UN related organizations are facing difficulties in carrying out their activities in the.

African Union response

The lead international actor on Darfur has been the African Union (AU). The AU has done as much as it could within the framework of its limited resources and mandate. In 2004 the AU established a small monitoring mission, the African Mission in Sudan (AMIS), in Darfur, consisting of some 60 monitors and 300 troops to protect them. In mid-2005, the number of troops increased to about 7,000. Under UN Security Council Resolution 1564, AMIS mandate is largely to serve as an observer mission. It does not have the mandate to go out and practically protect civilians. AMIS was the only external military force in Sudan's Darfur region until the arrival of the AU-UN Hybrid Mission in Darfur (UNAMID). The AU has failed in the resolution of the crisis through AMIS. It simply extended its timeframe for months and months without bothering to amend its ineffective mandate for protection of civilians and made no provision for more African troops to come with greater mobility and firepower on the ground in order to stabilize the situation in the crisis-hit region. Nonetheless, the AU strategy for dealing with the Darfur crisis primarily through the 7,000 troops with a weak mandate and limited resources, collapsed.

1. The N'djamena ceasefire

On April 2004, Chadian and AU mediators brokered a first ceasefire in N'djamena. The ceasefire was to be monitored by AU observers (AMIS). The agreement made little progress, in part because of the disorganization of the rebels. The Ndjamena Humanitarian Agreement was the basis for all subsequent diplomatic efforts on Darfur. It was a rushed agreement, which exists in two versions without an agreed text although it allowed the

African Union (AU) to dispatch ceasefire monitors, and subsequently a force to provide protection for those monitors and any civilians in the immediate vicinity of its operations, it contained no maps to enable the ceasefire monitors to do their job.[87]

On 16 November after another eight months of wrangling, the Security Council finally obtained Sudan's consent to the hybrid AU–UN force ,the UN–African Union Mission in Darfur (UNAMID), which was duly mandated in Resolution 1769 of 31 July 2007, with its mandate, structure, size and talks determined with reference to a joint UN–AU assessment

2. The Darfur Peace Agreement

After seven rounds of negotiations in Abuja under the auspices of an African Union mediation team supported by the UN49, the Darfur Peace Agreement (DPA) was signed in May 2006 by the NCP and one of the rebel groups, the SLA/ Minni. There were several factors that negatively impacted the process.[88] The DPA includes some power and wealth sharing provisions, but "did not address core issues that drive the conflict, such as land tenure and use, grazing rights, and the role and reform of local government and administrative structures.[89]

3. The roadmap and the Sirte talks

The AU/ UN mediation team strategy was initially to promote the implementation of the DPA, hoping that the non-signatories would eventually join. But they soon realized that a new negotiation process was needed, as the DPA "continues to be largely ignored by the parties on the ground".

---

85. Ibid.
86. Darfur's New Security Reality, ICG Africa Report N° 134" (International Crisis Group, November 2007), executive summary. Available at <http://www.ciaonet.org/wps/icg10310/icg10310.pdf>.[accessed on 25 July 2012].
87. Darfur's New Security Reality, ICG Africa Report N° 134," 22.

The first attempt was to simply expand the DPA, but a "long-term solution must deal with all the conflict's root causes and incorporate many views, not just of those with arms. In August 2006, all fighting groups were invited to Arusha/ Tanzania to unify their positions. Most of them refused to come. Another round of talks launched in Sirte/ Libya on 27 October 2007which whereas the major rebel groups boycotted the event.[90]

4. Doha Agreement

In the summer of 2008, an AU/ UN Joint Chief Mediator for Darfur, Mr Djibrill Bassolé, was appointed, to replace the two Special Envoys Jan Eliasson (for the UN) and Salim Ahmed Salim (for the AU), and help facilitate the process, he keeps on reaching out to the different rebel movements. The talks resulted in signing declaration of Goodwill, on February 17 2009 in Doha, Qatar, between the government of and JEM.

---

88. Ibid.

Conclusion

Almost ten years have passed since the crises in Darfur came into existence, during this period of time an intensive effort has been exerted and billions of dollars have been spent over negotiations, humanitarian aid and the deployment of military troops, nevertheless, the crisis is unfold but evolving. The involvement of the International community in Darfur is unprecedented, if the problem was political one it would have been solved by now. It's high time for International community to look at the crisis from another untraditional perspective, try to look deep for the root causes rather than the symptoms. The crisis in Darfur is not an accidental disaster, or a human catastrophe that humanitarian intervention can reverse or solve. The conflict is as not simple as presented in the media, which casts the conflict in terms of Arabs versus black Africans. The reality is far more muddled. It is a human tragedy, brewing for decades that finally erupted into a vicious cycle of violence in 2003.

The Darfur conflict is not only a problem for the Sudanese, but is also a regional problem. The conflict is threatening the stability in other regions of Sudan, and in neighboring countries like Chad, and the Central African Republic. Looking at its nature, magnitude and internal dynamics that fuel it, one can see that, unless it is addressed soon, the conflict has the potential to plague the whole region with a continuous cycle of violence and lawlessness.

It's regrettable that, this climate change conflict has been used by some states and non-state actor to generate political and financial gains mainly the United States and save Darfur coalition. Likewise, the western media which is motivated by politics and money participated in misleading the public opinion on the crises in Darfur. Much is being written about the ongoing genocidal war in Darfur, much less about the causes. The conflict is being presented in the media as a war between Arabs and Africans, with Arab militias, called *janjawid*, carrying out massacres, rape and pillage with the support of the Khartoum Government. This simplifies and misrepresents a very complex reality.

Recommendations

It has been proved that, political solution per-se is futile to solve the crises in Darfur. For a long term solution, the flowing recommendation should be incorporated in any proposed compromise by AU, UN or the stakeholders.

1. Increase adaptive capacity through improving environmental governance: by building the capacity of local environmental leaders to understanding climate change in Darfur, especially for drought management. The lack of drought management was a key issue in the conflict.
2. Land Tenure Reforms: Land tenure reforms will be critical in a post-conflict setting. Therefore, informal channels to initiate land discussion among camp inhabitants should be found. However, these dialogues must remain non-threatening to the state and national level institutions. Furthermore, due to the breadth and destruction of the conflict, the exact traditional *hawakeer* land tenure systems will not be easily re-introduced due to the changed socio-political setting in Darfur. The new land tenure systems must have the perceptions of equity, accessibility, and feasibility.
3. Strengthen Civil Society: climate change adaptation should be added to increase environmental awareness and foster local civil society institutions. "Social learning and adaptation include such collective activities as discourse, imitation, and conflict resolution."
4. Integrate climate change adaptation into livelihood programs: incorporating climate change adaptation strategies into livelihood trainings will increase awareness, improve environmental education, and incorporate indigenous knowledge in daily adaptation strategies.

5. Network with Environmental NGOs, Universities, and Humanitarian: for successful climate change adaptation and improving the social fabric. Also, non-governmental institutions, academics, environmentalists, and others to share knowledge and increase capacity of institutions for climate change and they can be used to encourage stakeholders to participate. Furthermore, organizing conferences can be an excellent opportunity for incorporating climate change into the wider Darfur conflict dialogue, especially for peace building programs.
6. Incorporate Climate Change into Peace Education: Introducing climate change education will help increase awareness of adults, but also expose children to the issue of global warming. As climate change becomes more significant in the region, people must be aware of the causes and learn practical solutions for adaptation.
7. Future Research Conflict management and mitigation programs should consider the impacts of climate change in their assessments. Also, future programs should use scientific climate change models and socio-economic and political information to predict high conflict risk areas. These methods can help prevent conflicts, as well as improving communities' ability for adaptation in a changing climate.
8. International community should initiates sustainable integrated rural development projects, guided by pro-poor policies and centered on improved natural resources management for increased sustainable agricultural productivity, improved rural infrastructure, reliable services including extension, credit and crop protection in addition to improved social service and better marketing channels by:
    - Enhancing natural resources management through strengthening information base, land reform, conservation measures and sustaining the natural pastures.
    - Raising agricultural productivity through improved farming systems- improved technologies.

- Increase resource use efficiency.
- Improving agricultural production systems by providing inputs and support agricultural services like credit.
- Invest in rural infrastructure – roads, power, and drinking water.
- Improve social services.
- Increase investment in agriculture.
- Improve marketing system.

Glossary

1. Abbala	camel herders
2. AMIS	African Union Mission in Sudan
3. Ansar	followers of Mahadi
4. Baggara	cattle herders
5. CPA	comprehensive peace agreement
6. Dar	tribal homeland
7. DPA	Darfur Peace Agreement
8. DUP	Democratic Unionist Party
9. ESPA	East Sudan Peace Agreement
10. FROLINAT	Front for the liberation of Chad
11. Hakura	land grant under fur sultanate, plural *hawakeer*
12. ICC	International Criminal Court
13. IMF	International Monetary Fund
14. IPCC	intergovernmental Panel for Climate Change
15. Janjawid	formed from the Arabic words for "man", "gun" and horse
16. JEM	Justice and Equality Movement
17. Khalifa	successor: the title of the successor of the Sudanese Mahadi
18. Mahdiya	The Mahdist revolution (1882-1885) mahdist state (1885-1898)
19. Marahil	Routs used by pastoralists in their pursue of water and grazing
20. Milik	freehold, tribes in Darfur consider the land as their own milik
21. NCP	National Congress Party
22. NIF	National Islamic Front
23. Sheikh	lowest –ranking tribal leader
24. SLA	Sudan Liberation Army, rebel group in Darfur

25. SPLA	Sudan People's Liberation Army, militia group in south sudan
26. SPLM	Sudan People's Liberation Movement
27. Sultan	Title of chief ruler in the seventeenth to twentieth centuries
28. TMC	Transitional Military Council
29. UNAMID	UN-African Union Mission in Darfur
30. UNSC	UN Security Council
31. UNEP	UN Environmental program
32. UNCCD	UN Convention to combat desertification
33. UNMIS	UN Mission in Sudan
34. Wadi	seasonal watercourse

List of figures

Bibliography

1. Abdul-Jalil, M. A. (2006). The Dynamics of Customary Land Tenure and Natural Resource Management in Darfur. Available at <www4.fao.org/cgi, [accessed on 18 July 2012].
2. Abu Abd' Allah Muhammad al-Idrisi in Nehemiah Levtzion and J. F. P. Hopkins, Corpus of Early Arabic Sources for West African History (Cambridge, 1981; republished Princeton, NJ: Marcus Wiener, 2000), p. 114-115; 124.
3. Agnew C.T & Chappell A. (1999). Drought in the Sahel. GeoJournal, 48, 299–311.Contribution of Working Group II to the third assessment report of IPCC. Cambridge University Press, Cambridge.
4. Alex Cobham, "Causes of conflict in Sudan: Testing the Black Book", Queen Elizabeth House Working Paper Number 121, University of Oxford, January 2005, p. 17.
5. Al-Maqrizi in Nehemiah Levztion and J. F. P. Hopkins, eds. and trans. Corpus of Early Arabic Sources for West African History (Cambridge, 1981, reprinted Princeton, NJ, 2000), pp 353-54.
6. Awad, M. H. (1971). The evolution of landownership in the Sudan. The Middle East Journal, 25(4), 212-228.
7. Babiker, Mustafa. 2001. Resource Competition and Conflict: Herder/Farmer or Pastoralism/Agriculture? In African Pastoralism: Conflict, Institutions and Government, eds. M. A. Mohamed Salih, Ton Dietz and Abdel-Ghaffar Mohamed Ahmed, 135−144. London: Pluto Press.
8. BBC (2008) "Country Profile: Sudan.
9. Bechtold, P. K. A History of Modern Sudan. Middle East Journal 2009, 63(1), 149 – 150.
10. B.H. Bell's "Note on tribal lands", August 10, 1930, enclosed with a letter from D. Newbold, Kordofan Governor, addressed to all District Commissioners, May 28,1933; El Obied (Kn.P./1/f.1/34).

11. Braukamper, Ulrich (2000). 'Management of Conflicts over Pastures and Fields among the Baggara Arabs of the Sudan Belt' in Nomadic Peoples journal, Volume 4 issue 1.
12. Colin H. Kahl, States, Scarcity, and Civil Strife in the Developing World, Princeton University Press.2008), p. 99.
13. Climate and land degradation report No, 989. Available at: < http:// www.wmo.int/page/thems/wmoprod/documents/wmo989.E.pdf
14. Darfur's New Security Reality, ICG Africa Report N° 134" (International Crisis Group, November 2007), executive summary. Available at <http://www.ciaonet.org/wps/icg10310/icg10310.pdf>.
15. De Waal Alex & A. H. Abdel Salam eds, The Phoenix State: Civil Society and the Future of Sudan, (Red Sea Press 200)1, p.45.
16. De Waal Alex, War in Darfur and search for peace, (Harvard University 2007), p.75.
17. De Waal, Sudan: the Turbulent State" in de Waal (ed) War in Darfur and the Search for Peace p. 20.
18. De Waal A. Famine that Kills: Darfur, Sudan, 1984-85, (Oxford University press 2005), p.191.
19. Dewal Alex, International Affairs journal vol 83, 2007 Journal Publishing Ltd/The Royal Institute of International Affairs
20. Dupont, Alan & Pearman, Graeme, 2006, "Heating up the planet: Climate change and security", Lowry Institute for International Policy, Paper 12, Sydney, p. 55.
21. Dyab, Ahmed I. Identity and Peace Issues in Sudan. Khartoum. Dar Azza Press 2007), p. 94.
22. eaquist, J. W., Hickler, T., Eklundh, L., Ardö, J., and Heumann, B. W.: Disentangling the effects of climate and people on Sahel vegetation dynamics, Biogeosciences Discuss., 5, 3045-3067, doi:10.5194/bgd-5-3045-2008, 2008.
23. ElNur, I. (et al) (1992) Displaced and Refugee Studies in the Sudan an Annotated Bibliography, University of Juba 1992), p.89.

24. Environmental Degradation Triggering Tensions and Conflict in Sudan (Nairobi: UNEP, 2007), Page 8.
25. Eastern Sudan peace agreement. Available at :< http://www.sudantribune.com/IMG/pdf/Eastern_Sudan Peace_agreement.pdf.
26. Fadul, A. A. Natural Resources Management for Sustainable Peace in Darfur. Sudan, University for Peace, (2004), p. 34.
27. Fadul, Abduljabbar Abdalla (2006) 'Natural resources management for sustainable peace in Darfur.' University for Peace: Addis Ababa.
28. Fadul, Abduljabbar A. 2004. Natural Resources Management for Sustainable Peace in Darfur. Paper presented at a Workshop organized by U. N. University for Peace & University of Khartoum Peace Research Centre on Environmental Degradation and Conflict in Darfur, December 15–16, in Khartoum, Sudan.
29. Failed states ranking. Available at www.fundforpeace.org.
30. Foreign policy briefing, volume No, 89 June 1, 2010.
31. Flint, Julie and Alex de Waal, Darfur: A Short History of a Long War, (Zed Books 2006), p. 25.
32. Frank Bajak, Chile-Haiti Earthquake Comparison: Chile Was More Prepared. The Huffington Post, February 27, 2010, http://www.huffingtonpost.com.
33. Gleditch, Nils Petter, Environmental Change, Security and Conflict. United States Institute of Peace Press 2007, pp.177-196; p. 179.
34. Homer-Dixon, T. F., 1994, 'Environmental Scarcities and Violent Conflict: Evidence from Cases', International Security, vol. 19, no. 1, pp. 5-40.
35. Homer-Dixon, Thomas. "Environmental Scarcities and Violent Conflict: Evidence from Cases," International Security, Vol. 19, No. 1 (summer, 1994), pp. 5-40; p. 10.
36. Homer-Dixon, environment, scarcity and violence, (Princeton University Press 2001), p.88.

37. IPCC third assessment report (2001). Climate Change Impacts, Adaptation and Vulnerability. Available at < http://www.ipcc.ch/publications_and_data/publications_and_data_reports International Organization for Migration. Available at < http://www.iom.int/jahia/page3.html,

38. John Barnett, The Meaning of Environmental Security: Ecological Politics and Policy in the new security era, (Zed books Ltd 2001), p.62.

39. Kahl States, Scarcity, and Civil Strife, p.108.

40. Kassa, A., 1999. Drought risk monitoring for the Sudan using NDVI. MSc Dissertation, University College, London. Available at http:/www.soas.ac.uk/Geography/Water Issues.

41. Land tenure and inter-ethnic conflict in Darfur. Paper presented at the Conference on Land Tenure and Conflict in Africa: Prevention, Mitigation and Reconstruction at the African Centre for Technology Studies (ACTS), December 9–10, 2004 in Nairobi, Kenya.

42. Michael Brzoska. Climate change as a driver of security policy, Paper prepare for SGIR 2010, (Stockholm: SGIR, September, 2010), pp 8-10.

43. Niloy Biswas, is the environment a security threat? Environmental Security beyond Securitization, international affairs review. Vol. XX, No. 1: Winter 2011. Available at www.iargwu.org.

44. O'Fahey, R.S. and Abu Salim M.I. Land in Dar Fur: Charters and Related Documents from the Dar Fur Sultanate. London: Cambridge University Press 1983.

45. Prunier, Gérard, Darfur: The Ambiguous Genocide, (Cornell University Press, 2005), pp. 8-24

46. Raleigh, C. and H. Urdal. "Climate change, environmental degradation and armed conflict." Political Geography Magazine 2007, 26(6): 674-694.

47. Rünger, M. Land Law & Land Use Control in Western Sudan: The case of southern Darfur. London: Ithaca Press 1987).p.5.

48. Security Council Resolution 1593 (2005).
49. Scott Straus, Darfur and the Genocide Debate, Foreign Affairs magazine, Vol. 84, No. 1 (Jan. – Feb., 2005), pp. 123/133.
50. Shaukat Hassan. Environmental Issues and Security in South Asia, ‖ ADELPHI Paper, no. 262, (Autumn 1991), pp 5-6.
51. Sjöström, M. (2004). Investigating vegetation changes in the African Sahel 1982-2002: a comparative analysis using Landsat, MODIS and AVHRR remote sensing data. Seminar series nr 106. Geobiosphere Science Centre and Physical Geography and Ecosystems Analysis. Lund University.
52. Stebbing E.P. The Creeping Desert in the Sudan and Elsewhere in Africa (McCorquodal & Co 1953), p.125.
53. Stern, N., (Ed.), the Economics of Climate Change: The Stern Review, Cambridge University Press, Cambridge, 2006, p. 3.
54. Suliman, M. 1998, The Inversion of Ethnicity from Perception to Cause of Violent Conflict, Friedensbericht, Ruegger Verlag, Zurich 1998),p.10.
55. Suliman, M. The conflict in Darfur: resource dimension. Available at <www.sudan-forall.org/.../Dr-Mohamed-Suliman.pdf>.
56. Suliman, M. The conflict in Darfur: resource dimension. Available at <www.sudan-forall.org/>
57. Tearfund. (2007). Darfur: water supply in a vulnerable environment p 17. Available at < http://postconflict.unep.ch/publications/UNEP_Sudan_Darfur_Water_Resources_TF.pdf
58. Teklu, T., Joachim von Braun, Zaki, E. (1991). Drought and Famine Relationships in Sudan:PolicyImplications.Available at:www.citeseerx.ist.psu.edu/viewdoc/download?doi=10.1.1.78>

59. Teklu, T., Joachim von Braun, Zaki, E. (1991). Drought and Famine Relationships in Sudan: Policy Implications.
60. United Nations Environment Programme, "Sudan Post-Conflict Environmental Assessment 2007, Page 93. Available at http://postconflict.unep.ch/publications/sudan/04_conflict.pdf.
61. UNEP, 2006. Africa Environment Outlook 2 – Our Environment, Our Wealth. Division of Early Warning and assessment (DEWA), United Nations Environment Programme (UNEP). Nairobi, Kenya.
62. UNEP Sudan post-conflict assessment report 2007.
63. UNEP: post conflict environmental assessment report, June 2007.
64. University for Peace, Environmental Degradation as a Cause of Conflict in Darfur (conference Proceeding), (Ethiopia: University for Peace, 2004), pp 12-13.
65. US Department of State (2004) www.state.gov/g/drl/rls/36028.htm.
66. UCDP (2008) Database www.ucdp.uu.se/database Uppsala University.
67. United Nations convention to combat desertification. Available at <http://www.unccd.int/en/regional-access/Africa/Pages/africa.aspx>
68. Wadi, A.I. Perspectives on Tribal Conflicts in the Sudan. (University of Khartoum 1998), p. 78.

About the author

Mohamed Osman Akasha is specialized in international security. He studied political science, and then pursued postgraduate courses in international relations at the Netherlands Institute of International Relations (Clingendael), the Canadian Institute for Conflict Resolution and Saint Paul University. He continued his research at the University of East Anglia where he focused on the untraditional concepts of international security. He served in the diplomatic corps as a diplomat for twelve years, three of them as a liaison officer with African Union mission in Darfur.

FSC
www.fsc.org
MIX
Papier | Fördert
gute Waldnutzung
FSC® C083411

Kolibri 360 GmbH
Ferdinand-Jühlke-Straße 7
99095 Erfurt, Deutschland
produktsicherheit@kolibri360.de